Recent Advances in
SUSTAINABLE PROCESS
Design and Optimization

Recent Advances in
SUSTAINABLE PROCESS
Design and Optimization

Contributors :
Siti H. Sarijo,
Zulkarnain Zainal, *et al.*

AURIS REFERENCE LTD.
London, UK

Recent Advances in Sustainable Process : Design and Optimization
Contributors : Siti H. Sarijo *and* Zulkarnain Zainal, *et al.*

Auris Reference Ltd., UK

www.aurisreference.com

United Kingdom

Copyright 2016

Printed in 2017 for Sale in the Indian Subcontinent

Notice

Recent Advances in Sustainable Process : Design and Optimization

ISBN: 978-1-78154-517-1

British Library Cataloguing in Publication Data
A CIP record for this book is available from the British Library

Exclusively distributed by CBS Publishers & Distributors Pvt. Ltd.

Sales & Distribution Rights only for India, Pakistan, Bangladesh, Sri Lanka, Nepal and Bhutan.This book is not to be sold outside these territories.

PREFACE

Sustainable Process Design in Industrial Chemistry is related to the entire life cycle of "things," from the development of substances and materials to their manufacturing process, waste disposal, and recycling. Basic education, laboratory works, and discussions are emphasized, and the related department aims to cultivate well-qualified researchers and chemical engineers through top level research activities.

Sustainable Industrial Chemistry covers optimising complex chemical processes in environmental management, industry and essential services such as water delivery. Sustainable Industrial chemistry requires a broad knowledge of both chemical engineering and chemistry.

Industrial chemists work to optimise complex chemical processes and are often solving problems at the cutting edge of environmental management in industry and essential services such as water delivery and pollution treatment. The advances in Green and Sustainable Chemistry have developed to help monitor, understand and limit the changes to our world due to the varying levels of chemicals in our environment.

The salient features of Recent Advances in Sustainable Process Design and Optimization have been adequately discussed and amply covered. This evolutionary process continues, and it is desirable that the type of training should keep pace with or even anticipate the changing conditions of the industry. The author has presented information which he considers to be of fundamental importance. Nevertheless, it is hoped that this will prove a valuable asset to those reading for college diplomas or university degrees and to all engaged in the technical branches of the dairy industry.

The present publication allows the reader to find information about the Process Design and Optimization techniques that are commonly used to provide information on chemical composition and structure. It is the fervent hope of author that this book would serve a useful purpose.

This page left intentionally blank.

CONTENTS

Mohd Zobir Hussein, Nor Shazlirah Shazlyn Abdul Rahman, Siti H. Sarijo
and Zulkarnain Zainal

This page left intentionally blank.

LIST OF CONTRIBUTORS

Mohd Zobir Hussein

Department of Chemistry, Faculty of Science, Universiti Putra Malaysia, Serdang, Selangor 43400, Malaysia; E-Mails: nsshazlyn@gmail.com (N.S.S.A.R.); zulkar@science.upm.edu.my (Z.Z.)

Nor Shazlirah Shazlyn Abdul Rahman

Department of Chemistry, Faculty of Science, Universiti Putra Malaysia, Serdang, Selangor 43400, Malaysia; E-Mails: nsshazlyn@gmail.com (N.S.S.A.R.); zulkar@science.upm.edu.my (Z.Z.)

Siti H. Sarijo

Faculty of Applied Science, Universiti Teknologi MARA (UiTM), Shah Alam, Selangor 40450, Malaysia; E-Mail: siti_halimah_404@yahoo.com

Zulkarnain Zainal

Department of Chemistry, Faculty of Science, Universiti Putra Malaysia, Serdang, Selangor 43400, Malaysia; E-Mails: nsshazlyn@gmail.com (N.S.S.A.R.); zulkar@science.upm.edu.my (Z.Z.)

and

Advanced Materials and Nanotechnology Laboratory, Institute of Advanced Technology (ITMA), Universiti Putra Malaysia, Serdang, Selangor 43400, Malaysia

This page left intentionally blank.

Chapter 1

SUSTAINABLE PROCESS DESIGN

GIVING PROCESS PLANTS THE CAPACITY TO ENDURE

Sustainability, by definition, is the capacity to endure. Specifically, sustainability involves finding ways to use resources so that those resources are not depleted or irreparably damaged.

Sustainable process design means implementing the principles of conservation into a production environment. Really, this is all about efficiency. The best way to conserve resources is to not use them in the first place. The best way to prevent harmful waste products from entering the environment is to not create them in the first place. How do you do that? You design a process that allows you to work more efficiently.

Innovative process design prioritizes efficiency. From the internal workings of one modular skid system to the smart design of an entire plant, process design optimization builds in efficiency at every level. That means your process requires fewer industrial resources to function at its peak level.

By taking in less, you produce less waste. Waste prevention is an increasingly important operational priority for modern process plants. Disposing of hazardous industrial byproducts carries substantial cost, and no modern company wants to be viewed as a polluter.

Adhering to the principles of sustainable design creates an opportunity to establish a better emotional connection with the public. People like companies that act in an environmentally responsible manner. They tend to trust them more, and are more likely to purchase their products. For many food and beverage producers, the increased trust and support more than justify the investment in sustainable design.

PRINCIPLES OF SUSTAINABLE DESIGN

The term "sustainable design" has been used in multiple disciplines, including but not limited to product design, architecture design, interior design, and graphic design. Sustainable design refers to the design process that integrates an environmentally friendly approach and considers nature resources as part of the design. Sharlyn Underwood, American Society of Interior Designers (ASID) Virginia chapter president and interior designer with SmithLewis Architecture, defines sustainable design in the architectural sector this way: "Sustainable design is the practice of designing buildings so that they exist in harmony with natural systems."

Sustainable design acts as a philosophy that is applied by different companies, governmental entities, and non-governmental organizations to achieve a better future for the human race through the wise and low-volume consumption of Earth's resources. Companies and governments that have advanced design strategies have more potential to apply sustainable design than others. Companies such as IKEA, for example, are taking advanced steps toward building sustainable products. Additionally, many governments that implement national design policies have provided positive steps toward applying sustainability.

The Role of Design in Sustainability

As designers, we have to understand our critical role in the sustainable world. One of the designer's roles is to solve problems and provide innovative solutions through products or services. Considering the critical problems that face our planet due to the irresponsible consumption of natural resources, designers play an important role in providing solutions to this problem and replacing obsolete products with innovative and sustainable ones that can ensure lower consumption of resources and less waste.

Looking at design through a holistic approach within the organization or company can unlock untapped capabilities to consider sustainable solutions that can replace old ones in existing products or envision new ones in the new product development (NPD) process. For example, the designers at LEGO contribute with the company team to create more planet-friendly materials for their toy bricks with the same properties and durability of the old bricks in order to maintain consumer satisfaction while upholding its planet promise.

Apple is another example of a company that depends on design and innovation to build an empathic relationship between it and its customers. Apple products have been improved over the years to become more sustainable. Electricity consumption during the sleep mode on the Apple iMac has improved 97 percent compared with the first iMac introduced in 1998. Additionally, the product design team has eliminated toxins from the whole design process.

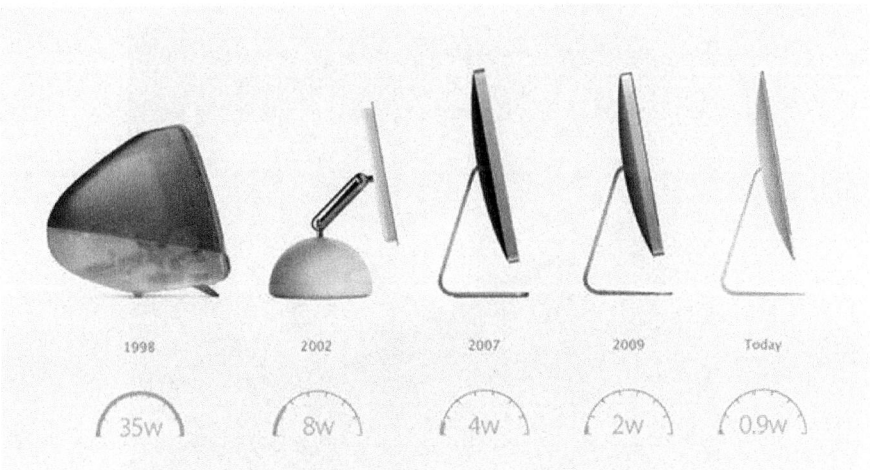

Fig. : Apple iMac evolution.

The designer's role starts with the very first design thoughts in the process of development and extends to the evaluation of the final product and analysis of consumers' feedback. Through every part of the process, sustainability should be considered wisely.

Principles of Sustainable Design

In order for the designer to consider sustainable design throughout the process, the question becomes what are the stages of the design process and design development that you can consider changing in order to make a more sustainable product? Below are some ideas on how to implement sustainability in design for each design area.

Form

The form represents the visual shape of the product and is usually perceived to be the main element of the design. Before designing a product's layout, however, the designer should ask questions like how will the shape affect energy consumption; and how will the size affect the packaging, transportation costs, and fuel emissions? IKEA's flat packing strategy, for example, helped it reduce transport costs, fuel usage, and emissions.

Function and Usability

The function and usability of the product contributes to its sustainability in an indirect way, as it helps consumers use the product more easily in less time and with less energy consumption. People do not want to keep hard-to-use products, so usable products can ensure less waste and throwaways.

Cost-Effective Solutions

Fig. : IKEA flat packing.

For many of today's sustainable products, cost is one of the key barriers that prevent many customers from making the switch from their dependence on non-sustainable products. Therefore, the designer and decision-makers are responsible for reducing the cost of current sustainable products.

Renewable Energy

Designers should stop depending on carbon energy and think in terms of building products that depend on renewable energy, such as solar panels and wind farms.

Materials and Recycling

Similar to energy, materials play an essential role in sustainable design, as every designer should search for materials that can be easily recycled or for which the planet can recreate in a short amount of time. For example, IKEA depends on mixed woods and innovative materials to replace traditional varieties of wood that can take a long time to grow in forests. The mixed and recycled materials can also help reduce product cost.

Durable Design Solutions

In order to reach zero waste, products have to either be durable enough to last for a long time or be fully recycled and transformed completely into new products. Depending on both methods can help recycle products more than one time and decrease the dependence on Earth's resources.

Constant Improvement and Sharing of KnowledgeEvaluation and improvement are important parts of any design process, but they take on even more importance in order to evaluate sustainable initiatives and improve them enough that they attain the same or better quality than existing products.

The principles above are general considerations that designers can depend on in order to build a sustainable design or service. Overall, the above design principles take into consideration the environment, people, economy, and culture. Every product or service design should consider these four factors. For example, the materials embedded in products should reflect concern for consumer safety and fit the cultural context in which they will be used.

Some obstacles that face some sustainable products are the result of a lack of consideration for these four values. In order to attract consumers to use a sustainable product, it should also address customer needs and compete in cost with other products on the market.

GUIDING PRINCIPLES OF SUSTAINABLE DESIGN

The term "guiding principle" has entered the contracting lexicon. Though vague, it means a benchmark for proper business practice. The term has established roots in environmental design. In fact, when talking about sustainable design, environmental design, green initiatives and the like, guiding principles are discussed quite often.

Sustainable Design

"Sustainable design is the practice of designing buildings so that they exist in harmony with natural systems," said Sharlyn Underwood, American Society of Interior Designers (ASID) Virginia Chapter president and interior designer with SmithLewis Architecture, Roanoke, Va. "Ideally, the resulting buildings contribute to human and ecosystem health while minimizing harm from their construction and operation. I actually prefer the terms 'green building' and 'high-performance design.'"

Sustainable design is an overall philosophy that embodies lower consumption, thus protecting the environment. Though the theory seems to be rooted in common sense, it is one that has struggled to become a common practice, which is why initiatives such as the Leadership in Energy and Environmental Design (LEED) from the U.S. Green Building Council (USGBC) has become so progressive and pervasive.

Sustainable design also has goals of creating healthier environments. Although more expensive than traditional design mechanisms, many are beginning to justify the additional costs associated based solely on the future energy savings. Guiding principles may not be an industry-wide accepted moniker because it may cause confusion.

"At this moment in time, there is no one set of guiding principles," said Underwood. "There are several frameworks from which to base sustainable design and gauge one's success at achieving it. The ultimate sustainable design principles are found in the cradle-to-cradle concept. It is a protocol developed by McDonough Braungart Design Chemistry that establishes guidelines for design to balance ecology, economy and environment.

"All products must be designed to be either infinitely recyclable in the technical product stream or be able to return to the natural environment with no harmful side effects," Underwood continued. "There are many other guidelines and principles available to individuals in the building industry including, but not limited to, USGBC's LEED Earthcraft house, Energy Star homes, and the National Association of Home Builders green guidelines."

Systems Take the Stage

A building's systems are prime candidates to benefit most from sustainable design. Lighting is perhaps the most discussed, as it has the ability to be environmental and energy efficient. Though the green aspects of lighting may seem to be counterintuitive to what electrical contractors sell, there is a place for harmony between the two elements.

"One of my charges lately has been to minimize exterior lighting and thus protect the dark sky. Protecting the dark sky is key when designing for exterior lighting in our parks. Emphasis is placed in lighting solutions that control glare, embrace appropriate lighting levels, are energy-efficient, minimize obtrusive light, display a good nighttime ambience, and have a minimal impact on artificial sky glow," said Ed Nieto, architect and illumination specialist, National Park Service, Denver. "These solutions and design techniques integrate full-shielded lighting that puts light where it is needed and minimizes glare and energy waste. Our goal is to provide excellent integration of lighting into overall design and promote versatile and distinctive dark-sky lighting solutions that enrich the park environment."

Nieto explains that within the National Park Service, current lighting schemes are almost reminiscent of those used at commercial venues. There has been a push through sustainable design to move toward a holistic lighting approach. "Exterior lighting in our National Parks primarily focuses on lighting for safety and security purposes," said Nieto.

This is a welcome change that is in line with the goal of protecting the night sky and stressing the importance of daylighting, two key components of green design relating to lighting systems. Nieto explained that projects are underway to move toward sustainability.

"A current lighting project that embraces responsible and sustainable practices is the lighting retrofit at Carlsbad Caverns. Our goal is to replace the existing lighting system with LED technology," he said. "This technology is very energy-efficient, addresses minimal resource impact with reduced relamping and provides excellent natural color rendering of the resource."

Generally speaking, most opt for using LEED as a guide for sustainable design. LEED has become a well-known design practice, and even though some may argue the true benefits, the majority agrees that most aspects are beneficial to energy conservation.

Getting on Board

On first glance, contractors may feel that sustainable design is out of their realm. But as Underwood explains, that is not the case. Since so many systems are elemental to a building's overall functionality, contractors are right up there at the top of the list of those that need to be in the know.

She said of the relationship between the various building systems are key to ensuring the building is running at maximum energy efficiency. In the LEED for new construction guidelines, controllability of lighting systems (both for individuals and overall building system), reduction of light pollution, and efficiency of light fixtures to optimize overall building energy performance all can help to achieve LEED credits," she said.

The best place for a contractor to start is to become more familiar with sustainable design.

"My number one resource of green building is the USGBC. ASID has been a member of the USGBC since their founding year and has greatly benefited from this exchange of information in leading the interior design industry in sustainable design," Underwood said.

Approach with Open Eyes

Sustainable design appears to have some solid roots. Due to mounting energy costs throughout the United States, many businesses and facilities have begun to look toward increased energy consumption. Though almost a double-edged sword, the rise in energy costs has helped bring sustainable design back to the forefront. Underwood believes that the future looks bright for going green, no pun intended.

"There is more and more documentation supporting green design. Research continues to be conducted on the cost savings, health benefits and environmental benefits of green design," she said. "As the documentation finds its way from the hands of designers and industry to the hands of the decision makers, it becomes difficult to challenge the benefits of green design and why one should not build green."

Underwood also describes how things have progressed. "When I jumped into green design about eight years ago, there were resources available and products emerging, but not as prevalent as they are today. It is challenging to pick up a building industry trade journal and not find a reference to green. In the last two to three years, I have seen a huge emergence of green as a topic in shelter magazines as well. Green products are becoming more readily available, but industry still needs a huge push to change," she said. "It can be challenging, but the benefits of going green are immense — the cost savings of health benefits and energy costs alone are enough to argue this case, not to mention the social benefits of designing, producing and building in a manner that will not inflict harm on our future generations. Those that are not leading the way in the industry today will be challenged to meet consumer demand as the consumer becomes more green savvy."

Between advances in products and technology, coupled with rising costs of building operation, going green continues to be a viable option for many entities. In fact, just the way the world has changed in general has helped make the case for sustainable design even stronger.

"In locations where governing bodies are not requiring or encouraging green building, the informed consumer will drive the demand for green buildings. This will impact the entire building community — it already has," she said. "As global warming continues to impact us — Hurricane Katrina, recent tornadoes in the Midwest, deluge of rain in California — our reliance on fossil fuels, practice of building within flood plains, and negligence of responsibility of design and products for their full life cycle will become more pronounced. Now is the time to think of future generations and our global environment. Green building is here is to stay and requires teamwork from the entire building industry."

Because of the nature of sustainable design, contractors should be aware and educated in its design practices and also the importance of the initiative. Other aspects of sustainability include water conservation, waste management, eco-friendly material usage, *etc.* Even though not all of the components of green are relevant to electrical contractors, there are enough critical elements that are. Especially with the trend toward integrated building systems will the lighting, energy and electricity portions become more important and instrumental that all contractors need be aware of what role they can and should play in this niche market.

BUILDING A SUSTAINABLE PROCESS INDUSTRY

The global chemical industry has always had the challenge of disposing of chemical wastes, by-products and residuals by legal, safe and economically effective means. Traditionally, the primary disposal methods have been landfill, deep well injection, and incineration. Responsible chemical companies all over the world have strived to dispose of these chemicals, especially hazardous ones, in keeping with national and state regulations. In hindsight, we have discovered that methods formerly thought to be safe and permanent have proven otherwise, as leaching, groundwater contamination, air quality issues, and other problems were discovered. Superfund legislation was enacted to identify sites needing remediation to avoid harm to the environment and to the public, and many of the sites identified met every environmental regulation at the time they were in use. Incineration, once believed to be the safest and most environmentally friendly way to dispose of chemical wastes, has its limitations as well.

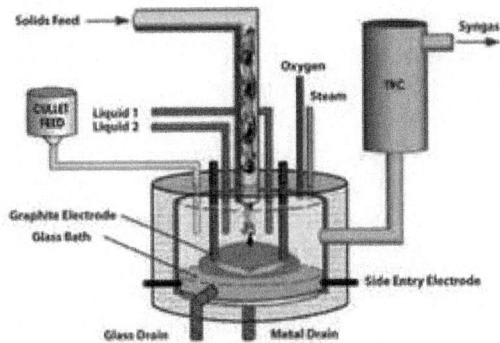

Fig. : Schematic of the G500 PEM system.

Historically, many chemical companies operated their own incinerators, but as emissions standards continued to tighten over the years, only very large incinerators could justify the high capital cost of installing appropriate emission controls. As a result, incineration capacity continues to fall, and conventional wisdom says no significant new hazardous waste incineration capacity will be built. Existing hazardous landfills are reaching capacity and permitting such new landfills has become difficult, very time consuming, and in many jurisdictions, impossible. Transportation costs for chemical residuals continue to rise and there are regulatory, environmental, and consumer pressures to reduce the movement of such products through public corridors. In addition, the chemical companies themselves have set goals to reduce their carbon footprints (the amount of CO_2 and other greenhouse gases produced by operations and transportation) and have established internal sustainability goals which seek to reduce the residuals produced in every process and to recycle and reuse as much material as possible.

Today, a new answer is commercially available to address all of these issues. The Plasma Enhanced Melter (PEMTM) gasification technology uses chemical wastes, by-products and residuals as feedstock and converts them into useful products returned to the chemical companies or sold to the broader market. PEM technology was invented by scientists and engineers working at the Battelle Pacific Northwest National Laboratory and the Massachusetts Institute of Technology, and commercialized by InEnTec LLC. These inventors found a unique way to combine a DC plasma arc with an AC glass melter to provide an efficient, cost-effective and environmentally sound process for converting waste materials into useful products.

Waste or chemical residuals (solid or liquid) are fed into the primary vessel, contacting the plasma zone which reaches temperatures of 3000°–10,000°C.

A plasma is a highly ionized gas, sometimes referred to as the fourth state of matter. Extreme examples of plasmas experienced in every day life are fluorescent lights and lightning. The PEM plasma is created by DC carbon arcs, which are consumed by the process and are automatically and continuously fed via a proprietary mechanism to the PEM. The glass bath is heated via AC resistance heating, just like a burner on an electric range. The combination of DC Plasma with an AC glass bath is unique and allows the energy to be balanced between the two zones in the most efficient manner. Oxygen and steam are fed into the PEM along with the waste at levels calculated, to provide reducing conditions such that high oxides cannot be formed. In short, within the PEM, a steam reforming reaction occurs almost instantaneously. Power requirements are generally proportional to the carbon content of the feedstock. The energy balance for typical wastes results in about one-third of the energy produced to be used to power the PEM and two-thirds are available for downstream use.

The refractory is proprietary and has shown excellent lifetimes, generally a minimum of three years, including exposure to high halogen containing wastes. Operation is highly automated, with numerous feedback loops and interlocks to ensure efficient and safe operation.

Organic materials are immediately converted to their elemental state and recombined to form synthesis gas. Most inorganic materials are dissolved or entrained in the glass. The glass can be collected on a batch or continuous basis and has been found to have useful properties. Metals may end up as part of the glass or in the case of heavier metals, pass through the glass to be collected separately. Halogens, sulphur, mercury, and other elements are removed from the synthesis gas in the gas treatment system, resulting in a very high purity synthesis gas that can be converted into a variety of products, such as hydrogen, methanol, ethanol, and dimethyl ether (DME).

The process has been subjected to numerous tests and has been found to meet or exceed all environmental regulations. Because it is not a combustion process, the PEM process is not conducive to the formation of dioxins, furans and other pollutants. A number of PEM units have been built and operated commercially. Today, a unit in Taiwan processes a mixture of industrial and medical wastes, generally operating 24/7. Another unit was sold in Japan and used to demonstrate PCB destruction for the Japanese government, and then relocated to demonstrate asbestos reduction. Based on those tests, an agency of the Japanese government has ordered two more units for PCB destruction.

Systems to be deployed for the chemical industry are supplied by InEnTec Chemical LLC, a joint venture of InEnTec LLC and Lakeside Energy LLC. Lakeside Energy has committed up to $150 million in equity funding to build PEM-based projects. Systems for the municipal waste industry are supplied by InEnTec Energy Solutions LLC, a wholly owned subsidiary of InEnTec.

The first on-site unit in the chemical industry in the US is currently under construction at Dow Corning Corporation's facility in Midland, Michigan. This project will use chlorosilane residuals as its feedstock and return HCl and synthesis gas to Dow Corning for use as a raw material and to displace natural gas usage. It is expected to be operational by the end this year. During 2008, using a portable PEM unit transportable on two semi-trailers, the versatility and utility of the PEM were demonstrated to a number of major chemical companies who brought samples of their hazardous waste to the temporary site. Each was provided detailed analysis showing degree of destruction and purity of the synthesis gas produced. InEnTec Chemical has several other projects in various stages of development with these and other major global chemical companies.

Use of the PEM in the chemical industry creates a true paradigm shift. Chemical wastes are no longer problems to be managed but are now the raw material for a new chemical process. In certain applications under US EPA regulations, if materials that would otherwise be hazardous waste are used as ingredients to produce other chemical products, they are excluded from RCRA regulation and are simply considered chemical feedstocks. Materials used as feedstock to the PEM process qualify for this exclusion.

A typical build, own, operate installation will be built on or contiguous to the chemical company plant site. Frequently the feedstocks are fed directly to the PEM and the new products are returned to be used by the host site. In this closed loop mode, emission sources are minimized and can be drastically reduced. In addition to useful products, other installations capture the energy value of the feedstocks as a 'waste to energy' installation, and, at the same time, reduce CO_2 and NO_x emissions exponentially. The PEM provides a new economic model as well. In return for a long-term contract to supply the raw material (waste or other residuals) and a long-term contract to take the products produced, a PEM system will be built at no capital cost to the chemical company.

Next-generation technologies such as PEM offer a new paradigm and a new vision for the chemical process industry. For the first time, a broadly applicable, economically viable and environmentally sound solution to the problem of chemical waste is commercially available. The PEM process is poised to become a major factor in helping the chemical industry move toward true sustainability.

Synthesis Gas to Energy or Clean Fuels

When wastes are introduced into the PEM, they are broken down by the extraordinarily high temperatures of the plasma zone into their basic molecules, which recombine to reform the elements present. The organic materials are converted into synthesis gas, a combination primarily consisting of hydrogen and carbon monoxide along with minor quantities of other elements present in the organic wastes. This mixture goes through a gas cleaning train and the resulting product is a very high quality synthesis gas.

Synthesis gas may be used directly as a fuel, usually replacing natural gas. Synthesis gas may be further processed in a standard PSA reaction to recover the hydrogen, useful in hydrogen fuel cells. It also can be converted via catalytic reactions to alcohols, including methanol or ethanol. Another option is conversion to dimethylether, which can be used as a direct substitute for propane. Finally, synthesis gas from the PEM, just like synthesis gas from a coal gasification process, can be converted into diesel fuel via a Fischer-Tropps reaction.

The synthesis gas from the PEM process is of exceptional purity compared to other gasification processes and is therefore useful in a broad variety of downstream processes that are sensitive to contaminants.

SUSTAINABLE PROCESS DESIGN FOR DESERT INDUSTRIAL FACILITIES

Most Middle East chemical / industrial facilities are located at seashore locations and are selected for high volume sea transport. Low tidal surges and absence of typhoons make this an optimal selection. Access to power and low cost energy is readily available.

Sea Water Access: As the primary cooling for most Middle Eastern industrial parks is a common cooling seawater network, the access to seawater is generally heavily regulated as a site utility. New users and expansion programs typically petition well in advance for access to seawater. Expect that a significant network water study must be prepared at the owners cost to ensure that new users do not degrade the overall common cooling seawater networks performance to all site users.

Sea Water Cooling: Cooling in desert regions is not a simple matter. Traditionally seawater for cooling refinery sized facilities is the norm because of the significant cooling loads required. Enormous titanium heat exchangers are used to provide a closed loop cooling to ensure no process contact with the returning seawater. The cooling is generally once through – non process contacting. The Seawater flow rates can be massive - looking at 4 to 6 rows of >120″ pipe is quite impressive.

The Ti exchanger systems are very costly, as the cooling is directly tied to surface area. The seawater heat exchanger approach is limited to the inlet temperature of the sea and the outlet seawater temperature is regulated.

Seawater cooling tower configurations are used to provide additional cooling of the seawater with very low cycles of evaporation; approximately 1.2-1.6. Closed loop Heat exchangers and closed cell cooling towers configurations are used.

Additional cooling beyond the approach limits of the seawater cooling tower mean temperature must be accomplished by chillers. Chillers are problematic as the cooling load may be enormous, and generally require cooling tower water for heat reject

As you can imagine there are significant metallurgical and NDA/DA testing programs underway from the largest oil and gas suppliers to ensure the reliability of the heat exchanger systems and seawater. User group forums are held annually where key learning are presented to show sustainable performance of these critical exchangers.

Freshwater Cooling Towers: The use of desalinated water cooling towers in replacement of seawater cooling are under consideration to eliminate the potential of accidental discharge of industrial chemicals directly into the seawater returns.

FW-CT can reduce costly MOC costs on exchangers, may eliminate a complete loop and the associated 10 degree approach, and possibly remove the need for a very costly chiller. A comparable system cost on a recent project showed the FW-CT CAP-EX was 33% less expensive in comparison to a similarly SW-CT system. The critical selection between the SW and FW CT's is related to the lack of WB/DB cooling that is available in the humid and hot summer months. Immediately after the sun goes down, most facilities experience a significant cooling capacity drop. Evaporation only FW-CT's are impacted however the elimination of a secondary loop may far exceed this limitation.

FW-CT Evaporation is significant in lieu of closed loop cooling and hence requires desalination of seawater in desert regions for makeup. Do not be surprised if the FW-CT makeup is > 60% of the total desalinated water needs.

FW-CT's blowdown has water chemicals for corrosion, scaling, heat transfer enhancement, and biological control. The chemicals used for chemical treatment must be addressed for disposal and recycle/reuse. Blow down is somewhat reduced by higher cycles of concentration, however freshwater blow-down must be addressed as a aqueous waste stream.

Dust Storms: When designing the cooling system, dust storms must be addressed. SW-CT have low cycles of concentration and properly designed basins and outlets that return the dust storm dust back to the ocean. FW-CT will require the removal of deposited dust.

A typical design basis dust storm is 24 hours, with $5000\mu g$ micrograms/m^3 of air, generating approximately 1200 kg/day of dust loading. *A surprise to non-regional designers is that available meteorological data indicates that 99% of particles are <10μ (microns) with peaks at 7μ and 2.5μ and that 50% of the particles are organic in nature.* This is drastically different from other parts of the world, and the removal is outside the typical deep bed sand filter range.

Some of the best information for dust storm data can be obtained from the US Air Force who issues regional information for its air equipment operations. It is a very thorough investigation by US Military of TSP, PM10, PM2.5 of 15 sites in Middle East, including specific data for Qatar and UEA. The report is 68 pages long and has much data and many tables. It has detailed average information compiled over one year including size and chemical analysis. In Qatar most of the dust is not SiO_2 as originally expected. It is high in Al, Mg, Ca, K and sulfur. Very fine dust between 0.5 and 2.5 microns exceeds 1% lead concentration. This could cause disposal issues with 1 micron filter cartridges. The results are very surprising and may complicate the design solution applied to protect equipment and heat exchangers from fouling.

Such loading exceeds typical removal techniques. The majority of the dust is essentially scrubbed from the air in the cooling tower. A typical design basis requires the filtration system to limits recirculating TSS during a dust storm to 30-35 mg/l, returning to <5 mg/l in 24 hrs.

More robust particulate removal must be included. Examples for integrated filtrations systems. Combinations of hydro-clones, multimedia pressurized sand filters, pleated cartridges, back washable filters, and cross flow or microfiltration filtration are classically employeed.

Organic fouling effects are amplified with the presence of colloidal dust particles. The organic material has an endless surface area for attachment and the fine dust transports the organic fouling throughout the system. Agglomeration of organic/inorganic matter forms a matrix for fouling deposits. As such low flow regions must be addressed; heat exchanger heads, cooling tower sumps *etc.* Cooling tower sump "basin sweeping" systems have been developed in the industry such as Lakos that jet the cooling tower sump systems to an outlet with appropriate filtration systems.

Red Tides: Red Tides, or Harmful Algal Blooms, are a growing concern in the region due to increased nutrient loads in the sea from on shore activities and thermal pollution. Intake of Red Tide seawater will have minimal impact on once through cooling systems. Based on the massive intake volume of common cooling seawater, red tide removal processes are not provided beyond standard intake screens. Annual Jellyfish migrations are more problematic in the intakes.

However Red Tides will need to be considered for desalinated water pro-duction. The necessary pre filtration to support reverse osmosis and vacuum evaporation differ, however both will suffer extensive fouling during a red tide event. Most desalination facilities would simply not operate during a Red tide event. In severe cases, red tides can generate air toxin and the evacuation of the shore area may be necessary.

Dissolved Air Flotation (DAF) appears to be the most viable and used red tide water treatment process. In many recent seawater RO plant bids, the owner frequently specifies DAF as mandatory equipment to be a part of the package.

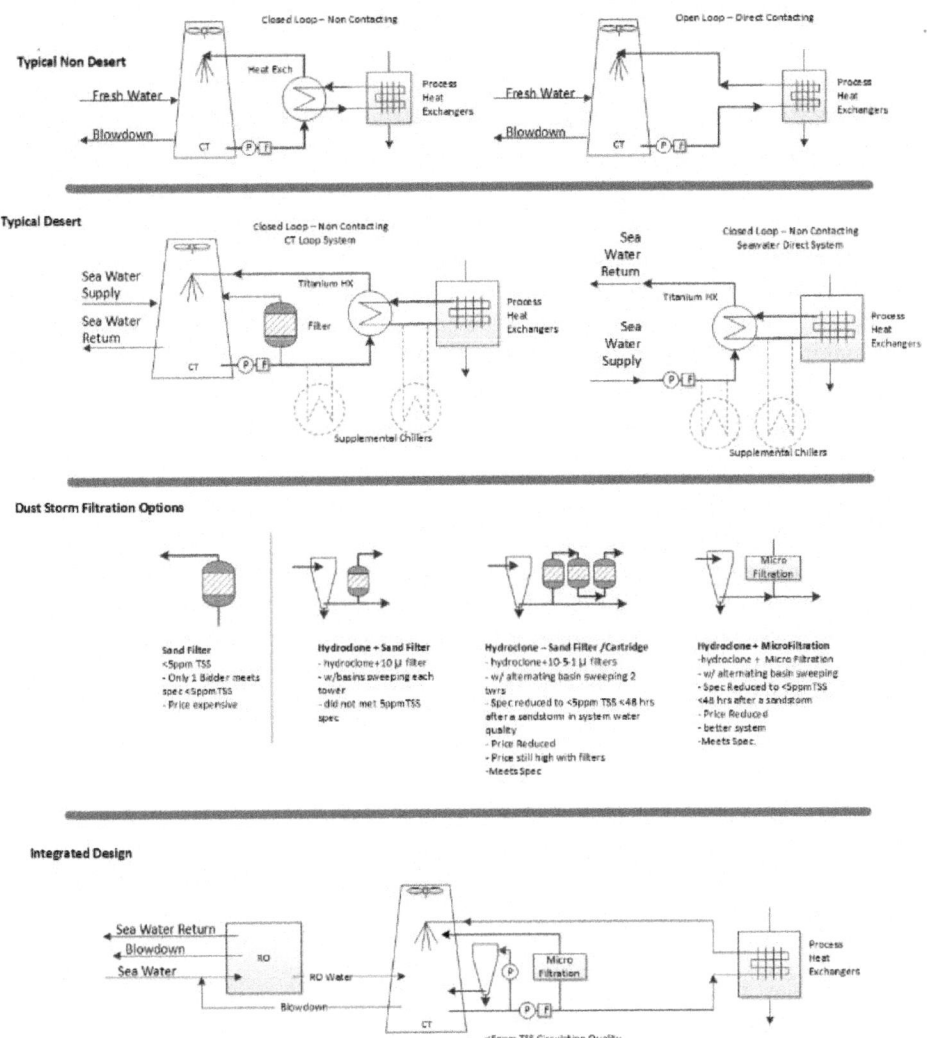

Sea Water Return: The Arabian Sea (GCC region) mean temperature and salinity has been steadily increasing. Virtually none of the coral reefs remain in the Arabian Sea due to temperature bleaching in the last decade. In Qatar for example new regulations are being considered / imposed to address thermal pollution impacts and their desire to protect the sea from chemical pollution. The goal is to ensure zero discharge of industrial waste or reclaimed waters to the sea water return.

Desalinated process and potable water: Desalinated water must be generated from Multi Stage Flash (MSF), Multi Effect Distillation (MED) with vapor recompression or Reverse Osmosis of seawater. The desalination of water generates a concentrated brine stream that is returned to the sea. Reverse Osmosis generates removes more minerals with less energy consumption and its brine is therefore more concentrated; the RO discharges approach most seawater return regional limits. In these industrial facility sea return outfalls the salinity and thermal impacts are highly regulated to minimize environmental impacts.

Process waste water and gray water: Desalinated and reclaimed waters used in the chemical facilities are evaporated or ultimately result in waste water. Although not expected, the humidity in these desert locations is similar to the US gulf coast but with higher summer design temperatures. Evaporation WB/DB cooling margin during peak summer months is surprisingly low. The disposal of waste water becomes a major issue. The energy costs to generate desalinated water necessitate its reuse. Recycle and reuse are standard programs in these facilities base on the financial considerations as well as environmental sustainability.

Produced gray water rates more closely matches the requirements for ground irrigation of green spaces.

Treated Industrial Water Paradigm: What makes desert conditions challenging is finding a sustainable outlet for the remaining treated industrial water. Most world regions have surface water – Owners discharge directly or via its Pretreatment plants, to POTW or Municipal Treatment Facilities, which discharge to rivers, and streams. Waste water is treated to drinking water standards or to a standard higher than the body of water it is being released. In most desert locations surface waters do not exist; what results is a very short list of alternatives.

Chapter 2

ADVANCED PROCESS CONTROL

Over the past 30 years, much have been written about advanced control; the underlying theory, implementation studies, statements about the benefits that its applications will bring and projections of future trends. During the 1960s, advanced control was taken to mean any algorithm or strategy that deviated from the classical three-term, Proportional-Integral-Derivative (PID), controller. The advent of process computers meant that algorithms that could not be realised using analog technology could now be applied. Feed forward control, multivariable control and optimal control philosophies became practicable alternatives. Indeed, the modern day proliferation of so called advanced control methodologies can only be attributed to the advances made in the electronics industry, especially in the development of low cost digital computational devices. Nowadays, advanced control is synonymous with the implementation of computer based technologies.

It has been recently reported that advanced control can improve product yield; reduce energy consumption; increase capacity; improve product quality and consistency; reduce product giveaway; increase responsiveness; improved process safety and reduce environmental emissions. By implementing advanced control, benefits ranging from 2% to 6% of operating costs have been quoted. These benefits are clearly enormous and are achieved by reducing process variability, hence allowing plants to be operated to their designed capacity.

What exactly is advanced control? Depending on an individual's background, advanced control may mean different things. It could be the implementation of feedforward or cascade control schemes; of time-delay compensators; of self-tuning or adaptive algorithms or of optimisation strategies. Here, the views of academics and practising engineers can differ significantly.

We prefer to regard advanced control as more than just the use of multi-processor computers or state-of-the-art software environments. Neither does it refer to the singular use of sophisticated control algorithms. It describes a practice which draws upon elements from many disciplines ranging from Control Engineering, Signal Processing, Statistics, Decision Theory, Artificial Intelligence to hardware and software engineering. Central to this philosophy is the requirement for an engineering appreciation of the problem, an understanding of process plant behaviour coupled with the judicious use of, not necessarily state-of-the art, control technologies.

This report restricts attention to control algorithms. Current approaches in this area rely heavily upon a study of system behaviour and the use of process models. Therefore this report will focus only on model based techniques. Although most of the methodologies to be described are applicable to a wide spectrum of systems, *e.g.* aerospace, robotics, radar tracking and vehicle guidance systems.

TYPES OF ADVANCED PROCESS CONTROL

Following is a list of the best known types of advanced process control:

- Advanced regulatory control (ARC) refers to several proven advanced control techniques, such as feedforward, override or adaptive gain. ARC is also a catch-all term used to refer to any customized or non-simple technique that does not fall into any other category. ARCs are typically implemented using function blocks or custom programming capabilities at the DCS level. In some cases, ARCs reside at the supervisory control computer level.

- Multivariable Model predictive control (MPC) is a popular technology, usually deployed on a supervisory control computer, that identifies important independent and dependent process variables and the dynamic relationships (models) between them, and uses matrix-math based control and optimization algorithms, to control multiple variables simultaneously. MPC has been a prominent part of APC ever since supervisory computers first brought the necessary computational capabilities to control systems in the 1980s.

- Inferential control: The concept behind inferentials is to calculate a stream property from readily available process measurements, such as temperature and pressure, that otherwise would require either an expensive and complicated online analyzer or periodic laboratory analysis. Inferentials can be utilized in place of actual online analyzers, whether for operator information, cascaded to base-layer process controllers, or multivariable controller CVs.

- Sequential control refers to dis-continuous time and event based automation sequences that occur within continuous processes. These may be implemented as a collection of time and logic function blocks, a custom algorithm, or using a formalized Sequential function chart methodology.

- Compressor control typically includes compressor anti-surge and performance control.

Related Technologies

The following technologies are related to APC and in some contexts can be considered part of APC.

- Statistical process control (SPC), despite its name, is much more common in discrete parts manufacturing and batch process control than in continuous process control. In SPC, "process" refers to the work and quality control process, rather than continuous process control.

- Batch process control is employed in non-continuous batch processes, such as many pharmaceuticals, chemicals, and foods.

- Simulation-based optimization incorporates dynamic or steady-state computer-based process simulation models to determine more optimal operating targets in real-time, *i.e.* on a periodic basis, ranging from hourly to daily. This is sometimes considered a part of APC, but in practice it is still an emerging technology and is more often part of MPO.

- Manufacturing planning and optimization (MPO) refers to ongoing business activity to arrive at optimal operating targets that are then implemented in the operating organization, either manually or in some cases automatically communicated to the process control system.

- Safety instrumented system refers to a system that is independent of the process control system, both physically and administratively, whose purpose is to assure basic safety of the process.

APC Business and Professionals

Those responsible for the design, implementation and maintenance of APC applications are often referred to as APC Engineers or Control Application Engineers. Usually their education is dependent upon the field of specialization. For example, in the process industries many APC Engineers have a chemical engineering background, combining process control and chemical processing expertise.

Most large operating facilities, such as oil refineries, employ a number of control system specialists and professionals, ranging from field instrumentation, regulatory control system (DCS and PLC), advanced process control, and control system network and security. Depending on facility size and circumstances, these personnel may have responsibilities across multiple areas, or be dedicated to each area. There are also many process control service companies that can be hired for support and services in each area.

PROCESS MODELS

Any description of a system could be considered to be a model of that system. Although the ability to encapsulate dynamic information is important, some analysis and design techniques require only steady-state information. Models allow the effects of time and space to be scaled, extraction of properties and hence simplification, to retain only those details relevant to the problem. The use of models therefore reduces the need for real experimentation and facilitates the achievement of many different purposes at reduced cost, risk and time.

In terms of control requirements, the model must contain information that enable prediction of the consequences of changing process operating conditions. Within this context, a model could either be a mathematical or statistical description of specific aspects of the process. It can also be in the form of qualitative descriptions of process behaviour. A non-exhaustive categorisation of model forms. Depending on the task, different model types will be employed.

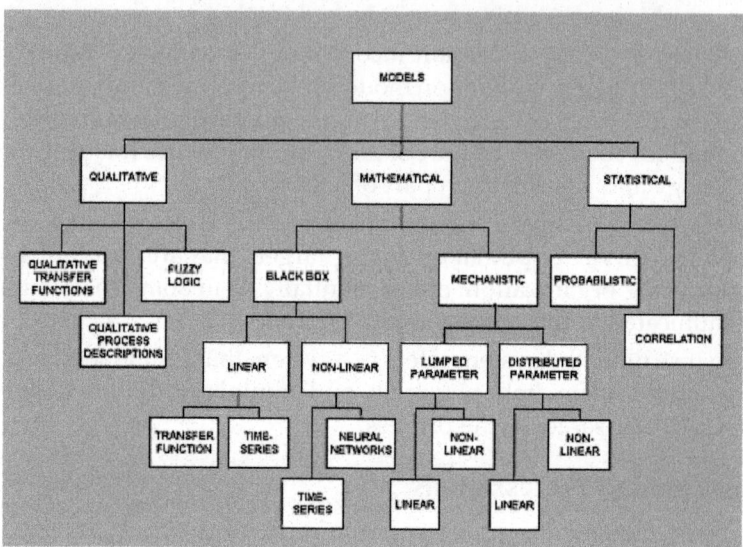

Fig. : Classification of Model Types for Process Monitoring and Control.

Mechanistic Models

If much is known about the process and its characteristics are well defined, then a set of differential equations can be used to describe its dynamic behaviour. This is known as 'mechanistic' model development. The mechanistic model is usually derived from the physics and chemistry governing the process. Depending on the system, the structure of the final model may either be a lumped parameter or a distributed parameter representation. Lumped parameter models are described by ordinary differential equations (ODEs) while distributed parameter systems representations require the use of partial differential equations (PDEs). ODEs are used to describe behaviour in one dimension, normally time, *e.g.* the level of liquid in a tank. PDE models arise due to dependence also on spatial locations, *e.g.* the temperature profile of liquid in a tank that is not well mixed.

Obviously, a distributed parameter model is more complex and hence harder to develop. More importantly, the solution of PDEs is also less straightforward. Nevertheless, a distributed model can be approximated by a series of ODEs given simplifying assumptions. Both lumped and distributed parameter models can be further classified into linear or nonlinear descriptions. Usually nonlinear, the differential equations are often linearised to enable tractable analysis.

In many cases, typically due to financial and time constraints, mechanistic model development may not be practically feasible. This is particularly true when knowledge about the process is initially vague or if the process is so complex that the resulting equations cannot be solved. Under such circumstances, empirical or 'black-box' models may be built using data collected from the plant.

Black Box Models

Black box models simply describe the functional relationships between system inputs and system outputs. They are, by implication, lumped parameter models. The parameters of these functions do not have any physical significance in terms of equivalence to process parameters such as heat or mass transfer coefficients, reaction kinetics, *etc.* This is the disadvantage of black box models compared to mechanistic models. However, if the aim is to merely represent faithfully some trends in process behaviour, then the black box modelling approach is just as effective. Moreover, the cost of modelling is orders of magnitude smaller than that associated with the development of mechanistic models.

The black box models can be further classified into linear and nonlinear forms. In the linear category, transfer function and time series models predominate. With sampled data systems, this delineation is, in a sense, arbitrary. The only distinguishing factor is that in time-series models, variables are treated as random variables. In the absence of random effects, the transfer function and time-series models are equivalent. Given the relevant data, a variety of techniques may be used to identify the parameters of linear black box models. The most common techniques used, though, are least-squares based algorithms.

Under the nonlinear category, time-series feature again together with neural network based models. In nonlinear time-series, the nonlinear behaviour of the process is modelled by combinations of weighted cross-products and powers of the variables used in the representation. The parameters of the functions are still linear and thus facilitates identification using least squares based techniques. Neural networks are not new paradigms to nonlinear systems modelling. However, the increase in cheap computing power and certain powerful theoretical results have led to a resurgence in the use of neural networks in model building.

Qualitative Models

There are instances where the nature of the process may preclude mathematical description, *e.g.* when the process is operated at distinct operating regions or when physical limits exist. This results in discontinuities that are not amenable to mathematical descriptions. In this case, qualitative models can be formulated. The simplest form of a qualitative model is the 'rule-based' model that makes use of 'IF-THEN-ELSE' constructs to describe process behaviour. These rules are elicited from human experts. Alternatively, Genetic Algorithms and Rule Induction techniques can be applied to process data to generate these describing rules. More sophisticated approaches make use of Qualitative Physics theory and its variants. These latter methods aim to rectify the disadvantages of purely rule based models by invoking some form of algebra so that the preciseness of mathematical modelling approaches could be achieved.

Of these, Qualitative Transfer Functions (QTFs) appear to be the most suitable for process monitoring and control applications. QTFs retain many of the qualities of quantitative transfer functions that describe the relationship between an input and an output variable, particularly the ability to embody temporal aspects of process behaviour. The technique was conceived for applications in the process control domain. Cast within an object framework, a model is built up of smaller sub-systems and connected together as in a directed graph. Each node in the graph represents a variable while the arcs that connect the nodes describe the influence or relationship between the nodes. Overall system behaviour is derived by traversing the graph, from input sources to output sinks.

Models derived based on the use of Fuzzy Set theory can also be classified as qualitative models. Proposed by Zadeh, fuzzy set theory contains an algebra and a set of linguistics that facilitates descriptions of complex and ill-defined systems. Magnitudes of changes are quantised as 'negative medium', 'positive large' and so on. The model combines elements of the rule based and probabilistic approaches and sets of symbols with interpretations such as, *'If the increment of the input is positive large, the possibility of the increment on the output being negative small is 0.8'*. Fuzzy models are being used in everyday life without our being aware of their presence, *e.g.* washing machines, auto focus cameras, *etc.*

Statistical Models

Describing processes in statistical terms is another modelling technique. Time-series analysis which has a heavy statistical bias may be considered to fall into this model category. Nevertheless, due to its widespread and interchangeable use in the development of deterministic as well as stochastic digital control algorithms, the earlier classification is more appropriate. The statistical approach is made necessary by the uncertainties surrounding some process systems. This technique has roots in statistical data analysis, information theory, games theory and the theory of decision systems.

Probabilistic models are characterised by the probability density functions of the variables. The most common is the normal distribution which provides information about the likelihood of a variable taking on certain values. Multivariate probability density functions can also be formulated but interpretation becomes difficult when more than two variables are considered. Correlation models arise by quantifying the degree of similarity between two variables by monitoring their variations. This is again quite a commonly used technique, and is implicit when associations between variables are analysed using regression techniques.

System dynamics are not captured by statistical models. However, in modern control practice, they play an important role particularly in assisting in higher level decision making, process monitoring, data analysis and obviously, in Statistical Process Control.

WHAT IS PROCESS MODELING?

Basic Definition

Process modeling is the concise description of the total variation in one quantity, y, by partitioning it into

1. A deterministic component given by a mathematical function of one or more other quantities, $x_1, x_2, ...,$ plus
2. A random component that follows a particular probability distribution.

Example

For example, the total variation of the measured pressure of a fixed amount of a gas in a tank can be described by partitioning the variability into its deterministic part, which is a function of the temperature of the gas, plus some left-over random error. Charles' Law states that the pressure of a gas is proportional to its temperature under the conditions described here, and in this case most of the variation will be deterministic. However, due to measurement error in the pressure gauge, the relationship will not be purely deterministic. The random errors cannot be characterized individually, but will follow some probability distribution that will describe the relative frequencies of occurrence of different-sized errors.

Graphical Interpretation

Using the example above, the definition of process modeling can be graphically depicted like this:

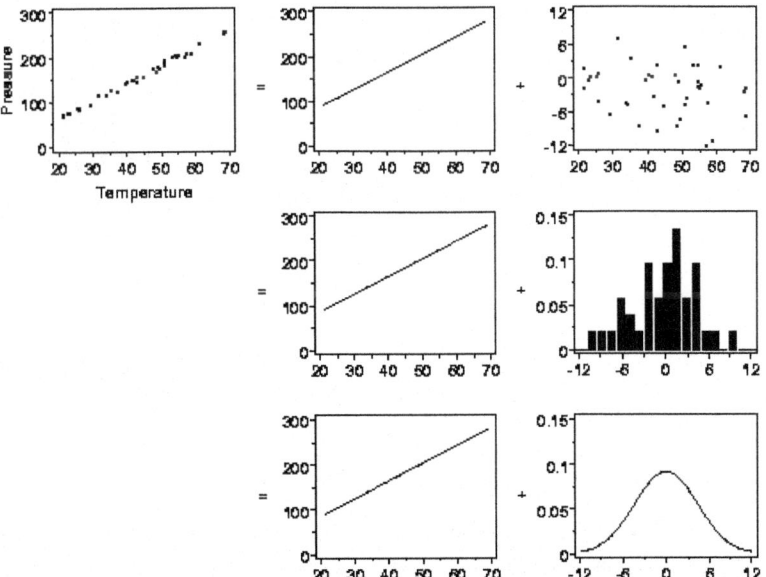

The top left plot in the figure shows pressure data that vary deterministically with temperature except for a small amount of random error. The relationship between pressure and temperature is a straight line, but not a perfect straight line. The top row plots on the right-hand side of the equals sign show a partitioning of the data into a perfect straight line and the remaining "unexplained" random variation in the data (note the different vertical scales of these plots). The plots in the middle row of the figure show the deterministic structure in the data again and a histogram of the random variation. The histogram shows the relative frequencies of observing different-sized random errors. The bottom row of the figure shows how the relative frequencies of the random errors can be summarized by a (normal) probability distribution.

An Example from a More Complex Process

Of course, the straight-line example is one of the simplest functions used for process modeling. Another example is shown below. The concept is identical to the straight-line example, but the structure in the data is more complex. The variation in *y* is partitioned into a deterministic part, which is a function of another variable, *x*, plus some left-over random variation. A probability distribution describes the leftover random variation.

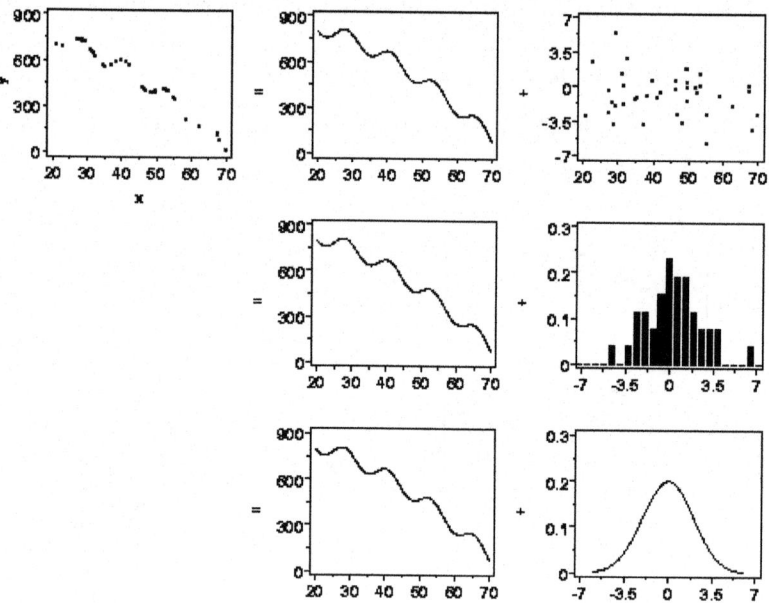

An Example with Multiple Explanatory Variables

The examples of process modeling shown above have only one explanatory variable but the concept easily extends to cases with more than one explanatory variable. The three-dimensional perspective plots below show an example with two explanatory variables. Examples with three or more explanatory variables are exactly analogous, but are difficult to show graphically.

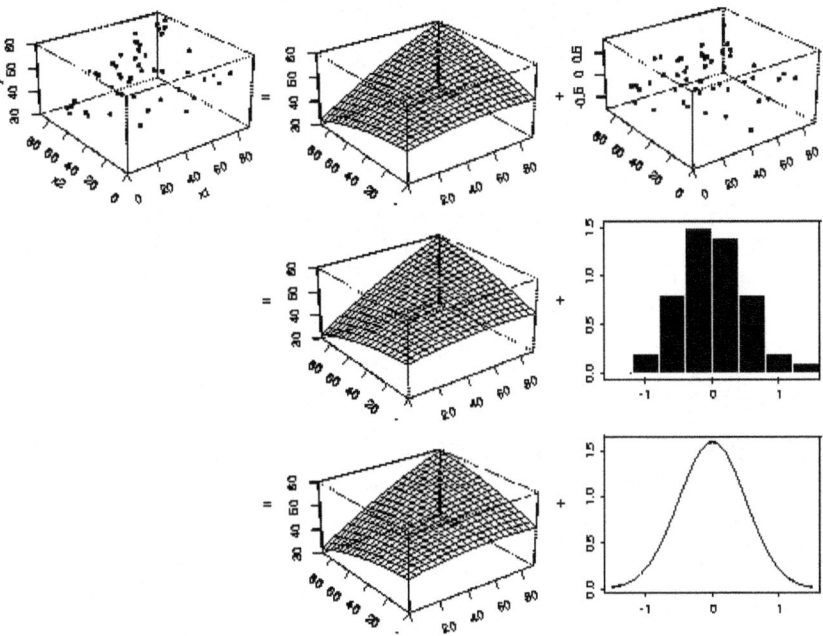

What Terminology do Statisticians use to Describe Process Models?

Model Components

There are three main parts to every process model. These are

1. The response variable, usually denoted by y,

2. The mathematical function, usually denoted as $f(\vec{x}; \vec{\beta})$, and

3. The random errors, usually denoted by ε.

Form of Model

The general form of the model is

$$y = f(\vec{x}; \vec{\beta}) + \varepsilon$$

The random errors that are included in the model make the relationship between the response variable and the predictor variables a "statistical" one, rather than a perfect deterministic one. This is because the functional relationship between the response and predictors holds only on average, not for each data point.

Some of the details about the different parts of the model, along with alternate terminology for the different components of the model.

Response Variable

The response variable, y, is a quantity that varies in a way that we hope to be able to summarize and exploit via the modeling process. Generally it is known that the variation of the response variable is systematically related to the values of one or more other variables before the modeling process is begun, although testing the existence and nature of this dependence is part of the modeling process itself.

Mathematical Function

The mathematical function consists of two parts. These parts are the predictor variables, $x_1, x_2, \ldots$, and the parameters, $\beta_0, \beta_1, \ldots$. The predictor variables are observed along with the response variable. They are the quantities described on the previous page as inputs to the mathematical function $f(\vec{x}; \vec{\beta})$. The collection of all of the predictor variables is denoted by $\vec{x}$ for short.

$$\vec{x} \equiv (x_1, x_2, \cdots)$$

The parameters are the quantities that will be estimated during the modeling process. Their true values are unknown and unknowable, except in simulation experiments. As for the predictor variables, the collection of all of the parameters is denoted by $\vec{\beta}$ for short.

$$\vec{\beta} \equiv (\beta_0, \beta_1, \cdots)$$

The parameters and predictor variables are combined in different forms to give the function used to describe the deterministic variation in the response variable. For a straight line with an unknown intercept and slope, for example, there are two parameters and one predictor variable

$$f(x; \vec{\beta}) = \beta_0 + \beta_1 x.$$

For a straight line with a known slope of one, but an unknown intercept, there would only be one parameter

$$f(x; \vec{\beta}) = \beta_0 + x.$$

For a quadratic surface with two predictor variables, there are six parameters for the full model.

$$f(\vec{x}; \vec{\beta}) = \beta_0 + \beta_1 x_1 + \beta_2 x_2 + \beta_{12} x_1 x_2 + \beta_{11} x_1^2 + \beta_{22} x_2^2$$

Random Error

Like the parameters in the mathematical function, the random errors are unknown. They are simply the difference between the data and the mathematical function. They are assumed to follow a particular probability distribution, however, which is used to describe their aggregate behavior. The probability distribution that describes the errors has a mean of zero and an unknown standard deviation, denoted by σ, that is another parameter in the model, like the β's.

Alternate Terminology

Unfortunately, there are no completely standardardized names for the parts of the model. Other publications or software may use different terminology. For example, another common name for the response variable is "dependent variable". The response variable is also simply called "the response" for short. Other names for the predictor variables include "explanatory variables", "independent variables", "predictors" and "regressors". The mathematical function used to describe the deterministic variation in the response variable is sometimes called the "regression function", the "regression equation", the "smoothing function", or the "smooth".

Scope of "Model"

In its correct usage, the term "model" refers to the equation above and also includes the underlying assumptions made about the probability distribution used to describe the variation of the random errors. Often, however, people will also use the term "model" when referring specifically to the mathematical function describing the deterministic variation in the data. Since the function is part of the model, the more limited usage is not wrong, but it is important to remember that the term "model" might refer to more than just the mathematical function.

What Are Process Models Used For?

Three Main Purposes

Process models are used for four main purposes:

1. estimation,
2. prediction,
3. calibration, and
4. optimization.

Estimation

The goal of estimation is to determine the value of the regression function (*i.e.*, the average value of the response variable), for a particular combination of the values of the predictor variables. Regression function values can be estimated for any combination of predictor variable values, including values for which no data have been measured or observed. Function values estimated for points within the observed space of predictor variable values are sometimes called interpolations. Estimation of regression function values for points outside the observed space of predictor variable values, called extrapolations, are sometimes necessary, but require caution.

Prediction

The goal of prediction is to determine either

1. The value of a new observation of the response variable, or

2. The values of a specified proportion of all future observations of the response variable for a particular combination of the values of the predictor variables. Predictions can be made for any combination of predictor variable values, including values for which no data have been measured or observed. As in the case of estimation, predictions made outside the observed space of predictor variable values are sometimes necessary, but require caution.

Calibration

The goal of calibration is to quantitatively relate measurements made using one measurement system to those of another measurement system. This is done so that measurements can be compared in common units or to tie results from a relative measurement method to absolute units.

Optimization

Optimization is performed to determine the values of process inputs that should be used to obtain the desired process output. Typical optimization goals might be to maximize the yield of a process, to minimize the processing time required to fabricate a product, or to hit a target product specification with minimum variation in order to maintain specified tolerances.

MODEL BASED (MODERN) AUTOMATIC CONTROL

Given a representative model of a process, 'What-If' investigations can be made via simulation, to answer operational questions such as safety related issues and to provide for operator training. However, this approach is not suitable for real-time automatic control. Within the context of automatic control, the inverse problem is considered, *i.e.* given the current states of the process, what actions should be taken to achieve desired specifications. Depending on the form of the plant model, different control strategies can be developed. The attraction of adopting a model based approach to controller development.

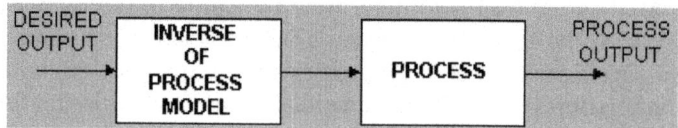

Fig. : Ideal Model Based Control.

By regarding the blocks to be mathematical operators, it can be seen that if an accurate model of the process is available, and if its inverse exists, then process dynamics can be cancelled by the inverse model. As a result, the output of the process will always be equal to the desired output. In other words, model based control design has the potential to provide perfect control. Hence, the first task in the implementation of modern control is to obtain a model of the process to be controlled. However, given that there are constraints on process operations; that all models will contain some degree of error and that all models may not be invertible, perfect control is very difficult to realise. These are the issues that modern control techniques aim to address, either directly or indirectly.

In the process industries, black box models are normally used for controller synthesis because the ill-defined nature of the processes makes mechanistic model development very costly. For process design purposes, precise characterisation is important. However, for the purposes of control strategy specification, controller design and control system analysis, models that can replicate the dynamic trends of the target processes are usually sufficient. Black box models have been found to be suitable in this respect and can be used to predict the results of taking certain actions.

Linear transfer functions and time-series descriptions are popular model forms used in control systems design. This is because of the wealth of knowledge that has been built up in linear systems theory. Increasingly, however, controllers are being designed using nonlinear time-series as well as neural network based models in recognition of the nonlinearities that pervade real world applications. The following sections briefly discuss the various algorithms that may arise from model based controller designs.

PID Control

The ubiquitous three-term Proportional+Integral+Derivative (PID) controller accounts for more than 80% of installed automatic feedback control devices in the process industries. In the past, these have been tuned using frequency response techniques or empirically derived rules-of-thumb. The modern approach is to determine the settings of the PID controller based upon a model of the process. The settings are chosen so that the controlled response adhere to user specifications. A typical criterion is that the controlled response should have a quarter decay ratio. Alternatively, it may be desired that the controlled response follow a defined trajectory or that the closed loop has certain stability properties.

It can be easily shown that a Proportional+Integral controller is optimal for a first order linear process without time-delays. Similarly, the PID controller is optimal for a second order linear process without time-delays. In practice, process characteristics are nonlinear and can change with time. Thus the linear model used for initial controller design may not be applicable when process conditions change or when the process is operated at another region.

One solution is to have a series of stored controller settings, each pertinent to a specific operating zone. Once it is detected that the operating regime has changed, the appropriate settings are switched in. This strategy, called parameter- or gain-scheduled control, has found favour in applications to processes where the operating regions are changed according to a preset and constant pattern. In applications to continuous systems, however, the technique is not so effective.

A more elegant technique is to implement the controller within an adaptive framework. Here the parameters of a linear model are updated regularly to reflect current process characteristics. These parameters are in turn used to calculate the settings of the controller.

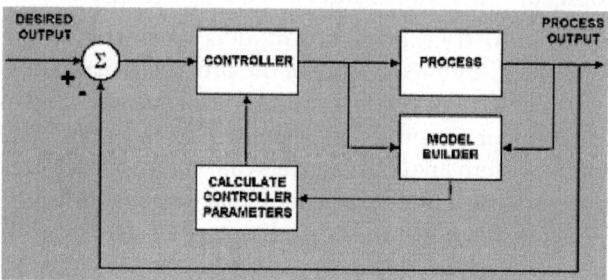

Fig. : Simplified Schematic of the Structure of Adaptive Controllers.

The settings of the controller can be updated continuously according to changes in process characteristics. Such devices are therefore called auto-tuning/ adaptive/self-tuning controllers. In some formulations, the controller settings are directly identified. A faster algorithm results because the model building stage has been avoided. Currently, many commercial auto-tuning PID controllers available from major control and instrumentation manufacturers. The simplest forms are those based upon the use of linear time-series models Some PID controllers are also auto-tuned using pattern recognition methods. For example, the Foxboro EXACT controller changes its settings to maintain a user defined response pattern. A good review of auto-tuning PID controllers is given in Astrom and Hagglund[1988].

Theoretically, all model based controllers can be operated in an adaptive mode. Nevertheless, there are instances when the adaptive mechanism may not be fast enough to capture changes in process characteristics due to system non-linearities. Under such circumstances, the use of a nonlinear model may be more appropriate for PID controller design. Nonlinear time-series, and recently neural networks, have been used in this context. A nonlinear PID controller may also be automatically tuned using an appropriate strategy, by posing the problem as an optimisation problem. This may be necessary when the nonlinear dynamics of the plant are time-varying. Again, the strategy is to make use of controller settings most appropriate to the current characteristics of the controlled process. A self-tuning PID controller based on the use of a nonlinear neural net model has been reported by Montague and Willis (1993).

Predictive Constrained Control

PID type controllers do not perform well when applied to systems with significant time-delays. Perhaps the best known technique for controlling systems with large time-delays is the 'Smith predictor'. It overcomes the debilitating problems of delayed feedback by using predicted future states of the output for control. Currently, some commercial controllers have Smith predictors as programmable blocks. There are, however, many other model based control strategies have dead-time compensation properties. If there is no time-delay, these algorithms usually collapse to the PID form. Predictive controllers can also be embedded within an adaptive framework and a typical adaptive predictive control structure.

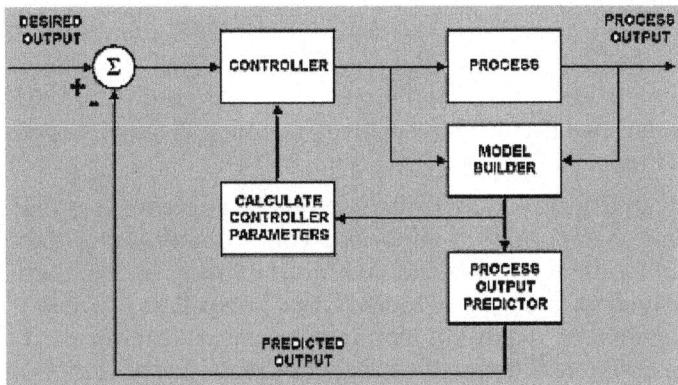

Fig. : Simplified Schematic of Adaptive Predictive Controllers.

The, by now, classical Generalised Minimum Variance (GMV) controller is an example of this philosophy. GMV control minimises the squared weighted difference between the desired value and the predicted output while penalising excessive control effort. The prediction horizon is the time-delay of the system, and this is a fixed parameter. GMV control, however, cannot effectively cope with variable time-delays and process constraints. This led to the development of long-range predictive controllers, *e.g.* the Generalised Predictive Controller (GPC) and Dynamic Matrix Control (DMC). The control problem is formulated in a manner similar to that adopted in the GMV approach. The differences are that the model is used to provide predictions of the output over a range of time-horizons into the future. Usually the range is between the smallest and largest expected delays. This alleviates the problem of varying time-delays and hence enhances robustness. Calculation of the control signal is essentially an optimisation problem. Here, economic objectives as well as process constraints can be included in the problem formulation. Examples of process constraints are the limits to liquid flows in fixed sized piping, allowable temperatures and pressures in process units, emissions to atmosphere, *etc.* Nowadays, the phrase 'predictive control' refers to the application of long-range predictive controllers. Again, predictive controllers may be designed using linear or nonlinear models.

Multivariable Control

Thus far, we have only considered the case where the is one manipulated input and one controlled output; single-input single-output (SISO) systems. With most processes, there are many variables that have to be regulated. The chemical reactor is a typical example where level, temperature and pressure have to kept at design values, that is there are at least three control loops; a multi-loop system. If the actions of one controller affect other loops in the system, then control-loop interaction is said to exist. If each controller has been individually tuned to provide maximum performance, then depending on the severity of the interactions system instability may occur when all the loops are closed. SISO controllers, whether adaptive, linear or nonlinear strategies, may therefore not be applicable to such processes. Models used in the design of SISO controllers do not contain information about the effects of loop interactions. Thus, they cannot be expected to perform well. For a multiloop strategy to work, individual SISO controllers are usually detuned (made less sensitive), resulting in sluggish performances for some or all loops.

Ideally, multivariable controllers should be applied to systems where interactions occur. As opposed to multi-loop control, multivariable controllers take into account loop interactions and their de-stabilising effects. Fortunately, it is a relatively trivial task to modify model based controllers to accommodate multivariable systems. By regarding loop interactions as feed-forward disturbances, they can be easily included in the model description. This simple augmentation leads to multivariable linear decoupling controllers, as well as nonlinear neural network based multivariable control algorithms. Following SISO designs, multivariable controllers that can provide time-delay compensation and handle process constraints can also be developed with relative ease. By incorporating suitable numerical procedures to build the model on-line, adaptive multivariable control strategies result.

Robust Control and the Internal Model Principle

Using an on-line parameter estimation algorithm to identify the parameters of the model, the parameters of most linear model based controllers can be adjusted in line with changes in process characteristics. Although great strides have been made in resolving the implementation issues of adaptive systems, for one reason or other, many practitioners are still not confident about the long term integrity of the adaptive mechanism. This concern has led to another contemporary topic in modern control engineering; robust control.

Robust control involves, firstly, quantifying the uncertainties or errors in a 'nominal' process model, due to nonlinear or time-varying process behaviour for example. If this can be accomplished, we essentially have a description of the process under all possible operating conditions. The next stage involves the design of a controller that will maintain stability as well as achieve specified performance over this range of operating conditions. A controller with this property is said to be 'robust'.

A sensitive controller is required to achieve performance objectives. Unfortunately, such a controller will also be sensitive to process uncertainties and hence suffer from stability problems. On the other hand, a controller that is insensitive to process uncertainties will have poorer performance characteristics in that controlled responses will be sluggish. The robust control problem is therefore formulated as a compromise between achieving performance and ensuring stability under assumed process uncertainties. Uncertainty descriptions are at best very conservative, whereupon performance objectives will have to be sacrificed. Moreover, the resulting optimisation problem is frequently not well posed. Thus, although robustness is a desirable property, and the theoretical developments and analysis tools are quite mature, application is hindered by the use of daunting mathematics and the lack of a suitable solution procedure.

Nevertheless, underpinning the design of robust controllers is the so called 'internal model' principle. It states that unless the control strategy contains, either explicitly or implicitly, a description of the controlled process, then either the performance or stability criterion, or both, will not be achieved. The corresponding 'internal model control' design procedure encapsulates this philosophy and provides for both perfect control and a mechanism to impart robust properties.

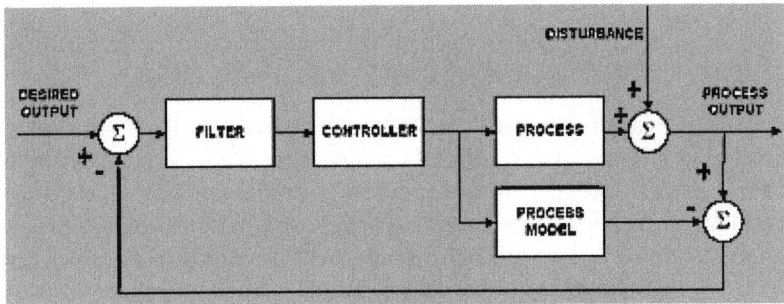

Fig. : Schematic of Internal Model Control Strategy.

If the process model is invertible, then the controller is simply the inverse of the model. If the model is accurate and there is no disturbance, then perfect control is achieved if the filter is not present. This also implies that if we know the behaviour of the process exactly, then feedback is not necessary! The primary role of the low-pass filter is to attenuate uncertainties in the feedback, generated by the difference between process and model outputs and serves to moderate excessive control effort. The strategy and the concept that it embraces are clearly very powerful. Indeed, the internal model principle is the essence of model based control and all model based controllers can be designed within its framework.

Globally Linearising Control

There are cases when adaptive linear control schemes would not perform well when faced with a highly nonlinear process. This is because the adaptive mechanism may not be fast enough to track changes in process characteristics. Appropriately designed nonlinear controllers would therefore be expected to perform better. The use of neural network model based controllers has already been mentioned. Another emerging field is that of nonlinear controller designed based on mechanisitc models via the use of differential geometric concepts. The aim of the design is similar to the use of Taylor series expansion to linearise the nonlinear model prior to application of linear model based controller designs. However, instead of providing local linearisation, contemporary nonlinear control strategies aim to provide 'global' linearisation over the space spanned by the states of the process; Globally Linearising Control (GLC). Global linearisation is achieved by a pre-compensator, designed such that the relationship between the inputs to the pre-compensator and the process output is linear. Linear control techniques can then be applied to the pseudo linear plant. A schematic of this strategy is shown in Figure below.

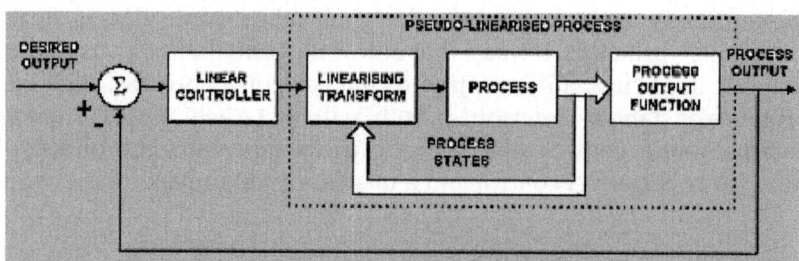

Fig. : Schematic of Globally Linearising Control.

Globally linearising control is a relatively new development and much research is being still being carried out to investigate the applicability of the technique. McLellan *et al* [1990] provide a review of nonlinear controller designs based upon mechanistic models. An interesting development that avoids the requirement of mechanistic models, is to use neural networks models instead. Neural network models are transformed into an equivalent state-space representation, and the GLC is designed based upon this state-space model.

MODEL PREDICTIVE CONTROL

Model predictive control (MPC) is an advanced method of process control that has been in use in the process industries in chemical plants and oil refineries since the 1980s. In recent years it has also been used in power system balancing models. Model predictive controllers rely on dynamic models of the process, most often linear empirical models obtained by system identification. The main advantage of MPC is the fact that it allows the current timeslot to be optimized, while keeping future timeslots in account. This is achieved by optimizing a finite time-horizon, but only implementing the current timeslot. MPC has the ability to anticipate future events and can take control actions accordingly. PID and LQR controllers do not have this predictive ability. MPC is nearly universally implemented as a digital control, although there is research into achieving faster response times with specially designed analog circuitry.

Overview

The models used in MPC are generally intended to represent the behavior of complex dynamical systems. The additional complexity of the MPC control algorithm is not generally needed to provide adequate control of simple systems, which are often controlled well by generic PID controllers. Common dynamic characteristics that are difficult for PID controllers include large time delays and high-order dynamics.

MPC models predict the change in the dependent variables of the modeled system that will be caused by changes in the independent variables. In a chemical process, independent variables that can be adjusted by the controller are often either the setpoints of regulatory PID controllers (pressure, flow, temperature, *etc.*) or the final control element (valves, dampers, *etc.*). Independent variables that cannot be adjusted by the controller are used as disturbances. Dependent variables in these processes are other measurements that represent either control objectives or process constraints.

MPC uses the current plant measurements, the current dynamic state of the process, the MPC models, and the process variable targets and limits to calculate future changes in the dependent variables. These changes are calculated to hold the dependent variables close to target while honoring constraints on both independent and dependent variables. The MPC typically sends out only the first change in each independent variable to be implemented, and repeats the calculation when the next change is required.

While many real processes are not linear, they can often be considered to be approximately linear over a small operating range. Linear MPC approaches are used in the majority of applications with the feedback mechanism of the MPC compensating for prediction errors due to structural mismatch between the model and the process. In model predictive controllers that consist only of linear models, the superposition principle of linear algebra enables the effect of changes in multiple independent variables to be added together to predict the response of the dependent variables. This simplifies the control problem to a series of direct matrix algebra calculations that are fast and robust.

When linear models are not sufficiently accurate to represent the real process nonlinearities, several approaches can be used. In some cases, the process variables can be transformed before and/or after the linear MPC model to reduce the nonlinearity. The process can be controlled with nonlinear MPC that uses a nonlinear model directly in the control application. The nonlinear model may be in the form of an empirical data fit (*e.g.* artificial neural networks) or a high-fidelity dynamic model based on fundamental mass and energy balances. The nonlinear model may be linearized to derive a Kalman filter or specify a model for linear MPC.

An algorithmic study by El-Gherwi, Budman, and El Kamel shows that utilizing a dual-mode approach can provide significant reduction in online computations while maintaining comparative performance to a non-altered implementation. The proposed algorithm solves N convex optimization problems in parallel based on exchange of information among controllers.

Theory Behind MPC

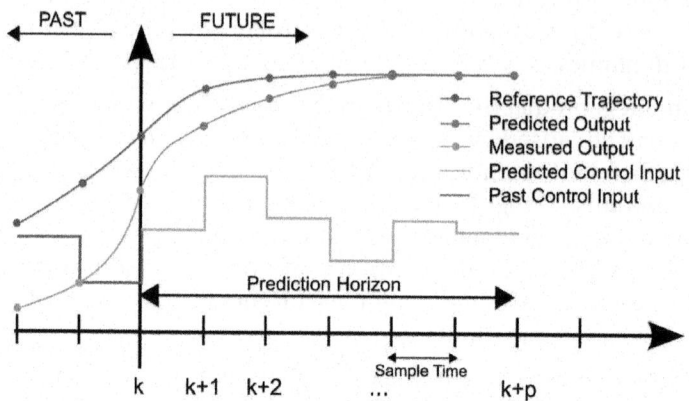

Fig. : A discrete MPC scheme.

MPC is based on iteratative, finite-horizon optimization of a plant model. At time t the current plant state is sampled and a cost minimizing control strategy is computed (via a numerical minimization algorithm) for a relatively short time horizon in the future: $[t, t + T]$. Specifically, an online or on-the-fly calculation is used to explore state trajectories that emanate from the current state and find (via the solution of Euler–Lagrange equations) a cost-minimizing control strategy until time $t + T$. Only the first step of the control strategy is implemented, then the plant state is sampled again and the calculations are repeated starting from the new current state, yielding a new control and new predicted state path. The prediction horizon keeps being shifted forward and for this reason MPC is also called receding horizon control. Although this approach is not optimal, in practice it has given very good results. Much academic research has been done to find fast methods of solution of Euler–Lagrange type equations, to understand the global stability properties of MPC's local optimization, and in general to improve the MPC method. To some extent the theoreticians have been trying to catch up with the control engineers when it comes to MPC.

Principles of MPC

Model Predictive Control (MPC) is a multivariable control algorithm that uses:

* an internal dynamic model of the process
* a history of past control moves and
* an optimization cost function J over the receding prediction horizon, to calculate the optimum control moves.

An example of a non-linear cost function for optimization is given by:

$$J = \sum_{i=1}^{N} w_{x_i} (r_i - x_i)^2 + \sum_{i=1}^{N} w_{u_i} \Delta u_i^2$$

without violating constraints (low/high limits)

With:

x_i = i -th controlled variable (*e.g.* measured temperature)

r_i = i -th reference variable (*e.g.* required temperature)

u_i = i -th manipulated variable (*e.g.* control valve)

w_{x_i} = weighting coefficient reflecting the relative importance of x_i

w_{u_i} = weighting coefficient penalizing relative big changes in u_i

etc.

Nonlinear MPC

Nonlinear Model Predictive Control, or NMPC, is a variant of model predictive control (MPC) that is characterized by the use of nonlinear system models in the prediction. As in linear MPC, NMPC requires the iterative solution of optimal control problems on a finite prediction horizon. While these problems are convex in linear MPC, in nonlinear MPC they are not convex anymore. This poses challenges for both NMPC stability theory and numerical solution.

The numerical solution of the NMPC optimal control problems is typically based on direct optimal control methods using Newton-type optimization schemes, in one of the variants: direct single shooting, direct multiple shooting methods, or direct collocation. NMPC algorithms typically exploit the fact that consecutive optimal control problems are similar to each other.

This allows to initialize the Newton-type solution procedure efficiently by a suitably shifted guess from the previously computed optimal solution, saving considerable amounts of computation time. The similarity of subsequent problems is even further exploited by path following algorithms (or "real-time iterations") that never attempt to iterate any optimization problem to convergence, but instead only take one iteration towards the solution of the most current NMPC problem, before proceeding to the next one, which is suitably initialized.

While NMPC applications have in the past been mostly used in the process and chemical industries with comparatively slow sampling rates, NMPC is more and more being applied to applications with high sampling rates, *e.g.*, in the automotive industry, or even when the states are distributed in space (Distributed parameter systems)

Robust MPC

Robust variants of Model Predictive Control (MPC) are able to account for set bounded disturbance while still ensuring state constraints are met. There are three main approaches to robust MPC:

- *Min-max MPC.* In this formulation, the optimization is performed with respect to all possible evolutions of the disturbance. This is the optimal solution to linear robust control problems, however it carries a high computational cost.

- *Constraint Tightening MPC.* Here the state constraints are enlarged by a given margin so that a trajectory can be guaranteed to be found under any evolution of disturbance.

- *Tube MPC.* This uses an independent nominal model of the system, and uses a feedback controller to ensure the actual state converges to the nominal state. The amount of separation required from the state constraints is determined by the robust positively invariant (RPI) set, which is the set of all possible state deviations that may be introduced by disturbance with the feedback controller.

Commercially Available MPC Software

Commercial MPC packages are available and typically contain tools for model identification and analysis, controller design and tuning, as well as controller performance evaluation.

CONTROL ENGINEERING

Fig. : Control systems play a critical role in space flight.

Control engineering or control systems engineering is the engineering discipline that applies control theory to design systems with desired behaviors. The practice uses sensors to measure the output performance of the device being controlled and those measurements can be used to give feedback to the input actuators that can make corrections toward desired performance. When a device is designed to perform without the need of human inputs for correction it is called automatic control (such as cruise control for regulating the speed of a car). Multi-disciplinary in nature, control systems engineering activities focus on implementation of control systems mainly derived by mathematical modeling of-systems of a diverse range.

Modern day control engineering is a relatively new field of study that gained significant attention during the 20th century with the advancement of technology. It can be broadly defined or classified as practical application of control theory. Control engineering has an essential role in a wide range of control systems, from simple household washing machines to high-performance F-16 fighter aircraft. It seeks to understand physical systems, using mathematical modeling, in terms of inputs, outputs and various components with different behaviors, use control systems design tools to develop controllers for those systems and implement controllers in physical systems employing available technology. A system can be mechanical, electrical, fluid, chemical, financial and even biological, and the mathematical modeling, analysis and controller design uses control theory in one or many of the time, frequency and complex-s domains, depending on the nature of the design problem.

History

Automatic control systems were first developed over two thousand years ago. The first feedback control device on record is thought to be the ancient Ktesibi-os's water clock in Alexandria, Egypt around the third century B.C. It kept time by regulating the water level in a vessel and, therefore, the water flow from that vessel. This certainly was a successful device as water clocks of similar design were still being made in Baghdad when the Mongols captured the city in 1258 A.D. A variety of automatic devices have been used over the centuries to accomplish useful tasks or simply to just entertain. The latter includes the automata, popular in Europe in the 17th and 18th centuries, featuring dancing figures that would repeat the same task over and over again; these automata are examples of open-loop control. Milestones among feedback, or "closed-loop" automatic control devices, include the temperature regulator of a furnace attributed to Drebbel, circa 1620, and the centrifugal flyball governor used for regulating the speed of steam engines by James Watt in 1788.

In his 1868 paper "On Governors", James Clerk Maxwell was able to explain instabilities exhibited by the flyball governor using differential equations to describe the control system. This demonstrated the importance and usefulness of mathematical models and methods in understanding complex phenomena, and signaled the beginning of mathematical control and systems theory. Elements of control theory had appeared earlier but not as dramatically and convincingly as in Maxwell's analysis.

Control theory made significant strides in the next 100 years. New mathematical techniques made it possible to control, more accurately, significantly more complex dynamical systems than the original flyball governor. These techniques include developments in optimal control in the 1950s and 1960s, followed by progress in stochastic, robust, adaptive and optimal control methods in the 1970s and 1980s. Applications of control methodology have helped make possible space travel and communication satellites, safer and more efficient aircraft, cleaner auto engines, cleaner and more efficient chemical processes.

Before it emerged as a unique discipline, control engineering was practiced as a part of mechanical engineering and control theory was studied as a part of electrical engineering since electrical circuits can often be easily described using control theory techniques. In the very first control relationships, a current output was represented with a voltage control input. However, not having proper technology to implement electrical control systems, designers left with the option of less efficient and slow responding mechanical systems. A very effective mechanical controller that is still widely used in some hydro plants is the governor. Later on, previous to modern power electronics, process control systems for industrial applications were devised by mechanical engineers using pneumatic and hydraulic control devices, many of which are still in use today.

Control Theory

There are two major divisions in control theory, namely, classical and modern, which have direct implications over the control engineering applications. The scope of classical control theory is limited to single-input and single-output (SISO) system design, except when analyzing for disturbance rejection using a second input. The system analysis is carried out in the time domain using differential equations, in the complex-s domain with the Laplace transform, or in the frequency domain by transforming from the complex-s domain. Many systems may be assumed to have a second order and single variable system response in the time domain. A controller designed using classical theory often requires on-site tuning due to incorrect design approximations.

Yet, due to the easier physical implementation of classical controller designs as compared to systems designed using modern control theory, these controllers are preferred in most industrial applications. The most common controllers designed using classical control theory are PID controllers. A less common implementation may include either or both a Lead or Lag filter. The ultimate end goal is to meet a requirements set typically provided in the time-domain called the Step response, or at times in the frequency domain called the Open-Loop response. The Step response characteristics applied in a specification are typically percent overshoot, settling time, *etc.* The Open-Loop response characteristics applied in a specification are typically Gain and Phase margin and bandwidth. These characteristics may be evaluated through simulation including a dynamic model of the system under control coupled with the compensation model.

In contrast, modern control theory is carried out in the state space, and can deal with multi-input and multi-output (MIMO) systems. This overcomes the limitations of classical control theory in more sophisticated design problems, such as fighter aircraft control, with the limitation that no frequency domain analysis is possible. In modern design, a system is represented to the greatest advantage as a set of decoupled first order differential equations defined using state variables. Nonlinear, multivariable, adaptive androbust control theories come under this division. Matrix methods are significantly limited for MIMO systems where linear independence cannot be assured in the relationship between inputs and outputs. Being fairly new, modern control theory has many areas yet to be explored. Scholars like Rudolf E. Kalman and Aleksandr Lyapunov are well-known among the people who have shaped modern control theory.

Control Systems

Control engineering is the engineering discipline that focuses on the modeling of a diverse range of dynamic systems (*e.g.* mechanical systems) and the design of controllersthat will cause these systems to behave in the desired manner. Although such controllers need not be electrical many are and hence control engineering is often viewed as a subfield of electrical engineering. However, the falling price of microprocessors is making the actual implementation of a control system essentially trivial. As a result, focus is shifting back to the mechanical and process engineering discipline, as intimate knowledge of the physical system being controlled is often desired.

Electrical circuits, digital signal processors and microcontrollers can all be used to implement control systems. Control engineering has a wide range of applications from the flight and propulsion systems of commercial airliners to the cruise control present in many modern automobiles.

In most of the cases, control engineers utilize feedback when designing control systems. This is often accomplished using a PID controller system. For example, in anautomobile with cruise control the vehicle's speed is continuously monitored and fed back to the system, which adjusts the motor's torque accordingly. Where there is regular feedback, control theory can be used to determine how the system responds to such feedback. In practically all such systems stability is important and control theory can help ensure stability is achieved.

Although feedback is an important aspect of control engineering, control engineers may also work on the control of systems without feedback. This is known as open loop control. A classic example of open loop control is a washing machine that runs through a pre-determined cycle without the use of sensors.

Control Engineering Education

At many universities, control engineering courses are taught in electrical and electronic engineering, mechatronics engineering, mechanical engineering, and aerospace engineering. In others, control engineering is connected to computer science, as most control techniques today are implemented through computers, often as embedded systems (as in the automotive field). The field of control within chemical engineering is often known as process control. It deals primarily with the control of variables in a chemical process in a plant. It is taught as part of the undergraduate curriculum of any chemical engineering program and employs many of the same principles in control engineering. Other engineering disciplines also overlap with control engineering as it can be applied to any system for which a suitable model can be derived. However, specialised control engineering departments do exist, for example, the Department of Automatic Control and Systems Engineering at the University of Sheffield and the Department of Systems Engineering at the United States Naval Academy.

Control engineering has diversified applications that include science, finance management, and even human behavior. Students of control engineering may start with a linear control system course dealing with the time and complex-s domain, which requires a thorough background in elementary mathematics and Laplace transform, called classical control theory. In linear control, the student does frequency and time domain analysis. Digital control and nonlinear control courses require Z transformation and algebra respectively, and could be said to complete a basic control education.

Recent Advancement

Originally, control engineering was all about continuous systems. Development of computer control tools posed a requirement of discrete control system engineering because the communications between the computer-based digital controller and the physical system are governed by a computer clock. The equivalent to Laplace transform in the discrete domain is the Z-transform. Today, many of the control systems are computer controlled and they consist of both digital and analog components.

Therefore, at the design stage either digital components are mapped into the continuous domain and the design is carried out in the continuous domain, or analog components are mapped into discrete domain and design is carried out there. The first of these two methods is more commonly encountered in practice because many industrial systems have many continuous systems components, including mechanical, fluid, biological and analog electrical components, with a few digital controllers.

Similarly, the design technique has progressed from paper-and-ruler based manual design to computer-aided design and now to computer-automated design or CAutoD which has been made possible by evolutionary computation. CAutoD can be applied not just to tuning a predefined control scheme, but also to controller structure optimisation, system identification and invention of novel control systems, based purely upon a performance requirement, independent of any specific control scheme.

Resilient Control Systems extends the traditional focus on addressing only plant disturbances to frameworks, architectures and methods that address multiple types of unexpected disturbance. In particular, adapting and transforming behaviors of the control system in response to malicious actors, abnormal failure modes, undesirable human action, *etc.* Development of resilience technologies require the involvement of multidisciplinary teams to holistically address the performance challenges.

STATISTICAL PROCESS CONTROL

Statistical Process Control (SPC) is widely applied in the parts manufacturing industries. Although the technique has been practised at various levels for more than 30 years, it warrants mention. In response to current total quality initiatives SPC has only just recently begun to be implemented in the process industries. SPC makes use of statistical models and procedures that to improve product quality and process productivity at reduced costs. The objective is to bring and keep processes in a state where any remaining variations are those inherent to the process.

Conventional SPC

SPC has been traditionally achieved by successive plotting and comparing a statistical measure of the variable with some user defined 'control' limits. If the plotted statistic exceeds these limits, the process is considered to be out of statistical control. Corrective action is then applied in the form of identification, elimination or compensation for the 'assignable' causes of variation. The most common charts used are the Shewhart, Exponential Moving Average (EWMA), range and Cumulative Sum (CuSum) charts.

Algorithmic SPC

Conventional SPC is basically an off-line technique. Whilst there are many reports of succesful cases in the parts manufacturing sector, this 'passive' control strategy does not suit continuous systems. Here, in addition to keeping products within specifications, there is a requirement to keep the process operating. Depending on the complexity of the process, the time taken to identify, eliminate and compensate for assignable causes of variation may not be acceptable. Nevertheless, the aim of both automatic process control and SPC is to increase plant profitability. Thus, it is reasonable to expect that the merger of these two apparently dichotomous methodologies could yield strategies that inherit the benefits associated with the parent approaches. This has been a subject of recent investigations where SPC is used to monitor the performances of automatic control loops. Such a strategy is sometimes called 'Algorithmic SPC' (ASPC), referring to the integrated use of algorithmic model based controllers and SPC techniques. Note, though, that the process is still being controlled by an automatic controller, that is the process is being controlled all the time.

Active SPC

Another way to integrate the two control approaches is to provide on-line SPC. Statistical models are used not only to define control limits, but also to develop control laws that suggest the degree of manipulation to maintain the process under statistical control. Thus, in applications to continuous processes, the need for an algorithmic automatic controller is avoided, leading to a direct or 'active' SPC strategy. Indeed, the technique is designed specifically for continuous systems. In contrast to ASPC, manipulations are made only when necessary, as indicated by detecting violation of control limits. As a result, compared to automatic control and ASPC, savings in the use of raw materials and utilities can be achieved using active SPC.

DEALING WITH DATA PROBLEMS

In the field of modern control engineering, much effort has been expended into the development and analysis of novel control strategies. A common assumption in these studies is that the required data is available. Unfortunately, this is often not the case in practice. However, the problem is gaining attention and a variety of solutions have been proposed to deal with various aspects of the difficulties associated with data.

Inferential Estimation

A major problem is the lack of on-line instrumentation to measure quantities that define product quality, *e.g.* stickiness of adhesives, smoothness of sheet material, melt flow index of polymers, flash points of fuels, *etc.* These are often provided by laboratory analyses resulting in infrequent feedback and substantial measurement delays, rendering automatic process control impossible. Inferential estimation is one method that has been designed to overcome this problem. The technique has also been called 'sensor-data fusion' and 'soft-sensing'.

Apart from the main quality variable, there are usually other variables such as temperatures, pressures, flows, *etc.*, that are associated with a process. Changes in some of these variables are indicative of changes in product quality. Thus, by monitoring suitable secondary variables, it is often possible to 'infer' the state of the quality variable. Process operators and engineers do this on a daily basis in running process plants. However, the process may be complex and there could be many factors that affect product quality. As a result, the relationship between process conditions and product quality may not be straight forward, leading to inaccuracies in human judgement. Inferential Estimation alleviates this problem. The technique uses easily obtainable measurements of variables that are known to influence product quality, together with those of product quality when available, to generate estimates of product quality.

As with feedback control strategies, in applications to non-linear systems, the relationship between secondary variables and the primary output can be 'learnt' automatically. Thus, parameters that define the relationship are adjusted to match changes in process characteristics. Alternatively, the inferential estimators may also be designed based upon the use of a neural network model. The estimates of the quality which are generated at the measurement frequency of the secondary variables may then be used for process monitoring or control purposes.

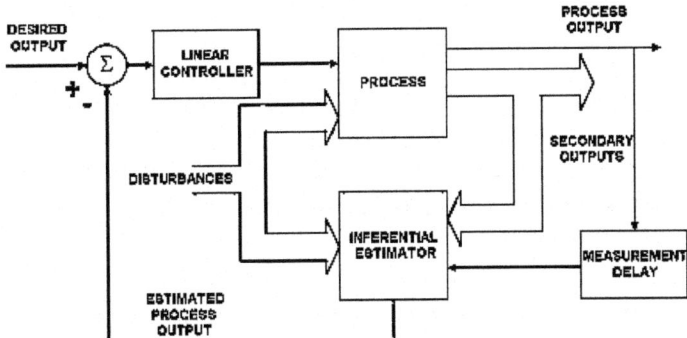

Fig. : Schematic of Feedback Control using Inferential Estimator.

Data Conditioning and Validation

Even if appropriate instruments exist, the data may not be of sufficient quality for desired goals to be achieved. Signals from plant are often corrupted by noise of varying magnitudes. All control methods are data driven. If appropriate measures are not taken to condition and validate the measured signals, then even the most sophisticated scheme will fail. In other words, the adage 'rubbish in, rubbish out' applies in the field of control.

In safety critical systems, such as the control and monitoring of nuclear reactors and power generators, steps are taken to ensure that the 'correct' signals are used for decision making. In these cases, it is common for both software and hardware redundancy schemes to be implemented. Redundancy is provided for by configuring software or hardware modules in duplicate or triplicate. Voting systems are then employed to validate output signals, retaining only those that are considered to be correct.

In less critical applications, duplex or triplex redundancy configurations are not cost effective. Therefore, unless there is absolute need, the smoothing of noisy signals is accomplished via hardware or software filtering to attenuate noise in measured signals. However, a penalty is incurred if the signal is subject to spikes. To remove these spikes or rogue points, heavy filtering has to be applied whereupon significant time-lags may be introduced into the filtered signal. Time-lags in the filtered signal may however be reduced by employing 'logic' filters which combine conventional filter algorithms with SPC concepts to validate and condition process measurements. This integrated approach has been shown to be very effective.

Data Analysis

Even if 'clean' data is available, there may be many variables associated with a particular process unit. The specification of an appropriate control strategy and controller design become complicated. Which variable should be manipulated to control another? What is the effect of this choice of manipulated-input controlled-output pairings? These are some of the important questions that have to be answered before a candidate configuration can be applied. Indeed, the results dictate the kind of models to employ for controller design and hence final controller types, and the overall control strategy that should be implemented. Inappropriate choice of input-output pairs exacerbate the problem of loop interactions. If interactions are significant, then a multivariable control design is necessary. If the input-output relations indicate nonlinear behaviour, then nonlinear controllers may have to be applied.

Many techniques can be used to tackle these issues. They range from simple graphical techniques (scatter plots, Box-plots), statistical multivariate analysis (*e.g.* Principal Component Analysis, Correlation Analysis, Cluster Analysis) to control theoretic, relative gain and singular value analyses. The latter are used to investigate control loop interactions and robustness of control strategies. However, there is currently no all embracing procedure for a systematic analysis of data right through to determining the suitability of the final control scheme.

HIGHER LEVEL OPERATIONS

Process Optimisation

The application of optimisation techniques is not restricted to the design of predictive constrained controllers. Process optimisation is a task in its own right. Unlike local controllers, which seek to maintain unit operating conditions at desired levels, the plant optimiser utilises a model of the plant to adjust operating conditions of the process so as to minimise raw material usage and maximise profits. The outputs of the optimiser therefore set the targets for the local controllers, taking into consideration the operational limits of the plant. This effectively bridging the gap between the plant's true business objectives and its actual operations. The generic configuration of a process optimisation scheme.

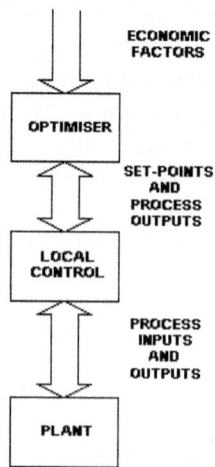

Fig. : Structure of Optimisation Scheme.

Due to the complexity and the scale of this type of optimisation problem, the model used is normally a steady-state description to enable a tractable solution. As with control algorithms, adaptive on-line optimisation is also feasible.

Process Monitoring, Fault Detection, Location and Diagnosis

Fault diagnosis has become an area of primary importance in modern process automation. It provides the pre-requisites for fault tolerance, reliability or security, which constitute fundamental design features in complex engineering systems. The system under consideration is monitored and the data is passed to fault detection algorithms or procedures. The basic task of a fault detection scheme is to register an alarm when an abnormal condition develops in the monitored system. Once a fault is detected, procedures may also be subsequently used to identify or diagnose the cause of the abnormality.

Fault detection and diagnosis techniques are again based upon the use of process models. In addition to the mathematical models used in controller design, statistical as well as qualitative models are increasingly being employed. Mathematical models are normally used to develop state-estimators or state-observers. Data from the monitored plant is input to these algorithms and the outputs compared with the corresponding plant outputs. If there are discrepancies, then it is an indication that at least one fault has occurred. The next task is to determine the locations of these faults. Again a representative model, not necessarily the one used in fault detection, is employed. In some instances, the location of the fault may be deduced by the type of fault. Here genetic algorithms and rule induction systems can be used to classify the fault.

Process Supervision via Artificial Intelligence Techniques

Human beings are able to make judgements in the face of subtle nuances and ambiguities. These knowledge processing capabilities cannot be matched by number crunching data processing algorithms. Although, the human decision system may not be precise, the result is often of sufficient accuracy for quick and effective problem solving. It has been the goal of computer scientists for many decades to build systems that mimic the decision making powers of human beings, *i.e.* artificial intelligent (AI) systems. AI techniques are also model based. Some would regard neural network based techniques to fall into the AI category. However, we tend to consider neural networks as numerical function approximators. Although AI techniques can make use of mathematical and statistical models, including neural networks, much of their utility is based upon the use of qualitative models.

Perhaps the most well known AI process supervisory schemes is based upon the use of expert systems. Expert systems are made up of three components. The rule or knowledge base holds information and logical rules for performing inference between facts. Next, there is the inference engine which controls the operation of the system and carries out the logical inference by processing the information in the knowledge base. The user interface makes up the final component, enabling communication between the user and the computer. Thus, an expert system is a collection of computer programs which operate upon the knowledge of experts in a particular application domain. Its purpose is to enable a novice to solve a problem with the benefits of the expert's knowledge.

When the inference engine and the user-interface are packaged as a single entity, this is known as an expert system shell. Software for procedures that can be combined together to form such a shell are known as expert system tools. The increasing availability of expert system shells and tools is a major reason for the proliferation of expert systems, where all that remains to be done is the compilation of the knowledge base. The extraction of rules that govern the operation of a process is called knowledge elicitation. This is performed via question and answer sessions between the extractor of knowledge, the so called knowledge engineer, and the provider of knowledge, the domain expert. There are also systems that are able to generate rules for expert systems when presented with data collected from a process. These are either based on rule induction techniques or genetic algorithms. However, the knowledge base could comprise any of the other qualitative models described previously and in any combination, including mathematical and statistical models.

When the system is presented with a collection of facts or a process scenario, the inference engine moves through the knowledge base in a 'forward' manner to come up with 'expert' advice or suggestions. However, unlike the implementation of 'IF-THEN-ELSE' constructs in conventional programming languages, expert systems have the ability to traverse the knowledge base in a backward direction. Backward chaining is invoked when the expert system is presented with a final result, and it is asked to provide a line of reasoning as to the events that led to the given result. Thus, another distinguishing factor of expert systems is that they are also able to provide explanations as to why a particular piece of advice or suggestion has been made. Expert systems have therefore found use in providing operator advice and as a process simulator for operator training.

Expert systems can be operated in two ways. The most common is the consultative mode where the expert system asks the user a series of questions. Alternatively, the data required by the expert system is provided directly by interfacing to plant instruments. There is a growing number of expert system shells that can reason in real-time. Such Real-Time Knowledge Based Systems (RTKBS) have been used to tune controllers, supervise the performance of adaptive controllers, perform fault detection and diagnosis, perform alarm management and even provide direct on-line process control.

ADVANCED CONTROL

The techniques have been applied to a wide variety of systems. In the process industries, they have been applied to reactors, separation processes, power generation systems including boilers, HVAC and so on. Many of these are reported by academics, academics involved in industrial collaborative projects or by consultants. There are also many unreported cases of successful advanced control applications, primarily because of commercial confidentiality. Many of the applications reported in the literature describe the use of single techniques. However, our philosophy of advanced control is depicted in the following diagram.

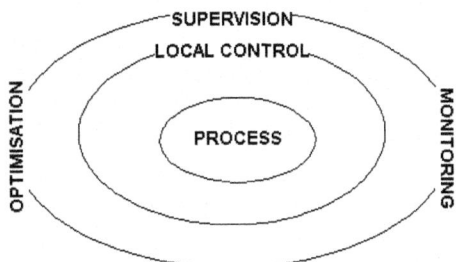

Fig. : Hierarchical Layers in Integrated Modern Control.

Local control is implemented, using appropriate controllers, to keep the process operating at desired conditions. Here, the type of local controllers employed depends on the task at hand. Although it is easier to tune and maintain simple controllers, some processes do require control by more sophisticated algorithms. However, unless such sophisticated controllers are installed and maintained by well trained trained personnel, they can be prone to failure. Until the last decade, the higher level tasks of monitoring, optimisation, and supervision were mainly carried out by human beings. Due to the advent of modern technology, and advances in the field of AI, these can now be automated. In particular, the installation, operation and integrity of modern controllers can be supervised by higher level systems.

Advanced control is the implementation of this hierarchical information and control structure. The flow of information is bi-directional, from management layer to process level and vice versa. The task here is to be able to integrate the various components in an efficient and manageable fashion. This can be facilitated by ensuring that each component is designed as a modular, yet integrable element.

CURRENT RESEARCH AND FUTURE TRENDS

In the process industries, the biggest challenge facing process engineers will be the reduction of variable costs whilst maintaining product quality. Advanced process control is the most effective technology available to realise this objective, especially on established plants. As systems become more complex, another important aspect is the reliability of the implemented systems. Here, the reliability of hardware and software are issues which have to be addressed. Allied to this is the requirement for suitably designed man-machine interfaces to enable efficient and reliable information transfer and to facilitate systems management.

With regard to the primary modules making up an advanced control project, neural networks, nonlinear systems theory, robust control, knowledge based systems are areas which appear to have captured the attention of both researchers and practitioners in the field of control engineering. This trend will continue well into the next decade. Areas that will receive particular attention will be techniques that will translate raw data into useful information; improved measurement methods including inferential estimation; multivariable non-linear predictive control and formal techniques for analysing the integrity of neural network based methodologies.

All information is of value, and should not be discarded just because they do not conform to a particular model building procedure. Thus, new modelling methods are also required. These should provide a framework where *a priori* knowledge of the process could be combined with the various existing modelling techniques, leading to so called 'grey-box' models. The resulting models should also be amenable for utilisation by the different modern controller designs, thus rendering controller synthesis independent of model types.

The process industries have an enormous base of manufacturing facilities which are still being run by unsophisticated or primitive control schemes. Competitive pressures will not allow any company in these industries to ignore the significant efficiencies possible through adopting modern process control technologies. A major obstacle to realising the full potential of modern control techniques is the lack of exposure to advances in the field. This can be overcome by the development of portable computer based training packages. The current proliferation in multi-media computing systems is the ideal impetus for the development of such learning aids.

Chapter 3

THE FOUNDATION FOR SUSTAINABILITY ASSESSMENT IN PROCESS DESIGN

DESIGN EDUCATION: LAYING THE FOUNDATION FOR SUSTAINABILITY

Graphic design education is missing a critical component. Students receive intensive training in many areas, such as foundation, design history, typography, and composition. But where does environmental sustainability fit on this list? At what point on our path as designers do we learn how to be better stewards of our environment?

The topic of sustainability in traditional design education is consistently an afterthought, if thought about at all. When sustainability is thought of, it usually takes the form of a single project set apart from real world expectations and limitations. These projects seem to be designed in a bubble as if students may only take environmental sustainability into consideration when it is specifically assigned.

There is a glimmer of hope as more design educators are changing their approach. Programs and degrees devoted to sustainable design are being offered at higher education institutions around the world. For example, Savannah College of Art and Design offers a master's degree in Sustainable Graphic Design, Minneapolis College of Art and Design offers a Sustainable Design certificate online, and Kingston University in London offers an MA by Research in Sustainable Design.

These new programs are a giant step in the right direction, but they are still segregated from typical graphic design programs in higher education. Currently, a student must seek out sustainable design programs, rather than learning about sustainability as part of her every day curriculum. This makes sustainability a separate type of graphic design practice when it should be as vital as every other design element such as kerning or information hierarchy.

Until all institutions teach sustainable graphic design as *the* way to design, students will not develop the skills, ability, or knowledge to lessen their environmental footprint. Young graphic designers will only continue an untenable cycle of materials waste, disengagement from supply chain practices, and disavowal of their own human and environmental impacts once they enter the workforce unless taught how to incorporate sustainability as the foundation to design.

When I started teaching at the New England School of Art and Design (NE-SAD), the school did not incorporate sustainability into its design curriculum. In my endeavor to educate students that environmentally friendly design is much more diverse and important than a requirement in a single isolated project, I introduced my *Sustainability & Ethics in Graphic Design* course. The goal was to teach students to make sustainability the foundation of *all* their work. In this seminar, students learn to apply sustainability principles and pragmatic thinking, gaining new insight into the design process and the relationships between humans and nature. *Sustainability & Ethics in Graphic Design* provides a context for the consideration of design beyond traditional models of practice.

This approach was successful, I believe, in laying the foundation of sustainable graphic design practice for students at NESAD. Lifecycle assessment, design planning and process, materials evaluation, and client education were covered with active discussions, individual and group projects, readings, written assignments, and guest speakers.

The first project required students to examine their own consumer behavior by conducting a self eco-audit. For three sets of three consecutive days, students studied their own environmental impact by collecting, noting, or photographing every piece of plastic, metal, and paper they consumed. This project enabled them to become familiar with their own consumption, use, and disposal before moving on to global issues. The students gained a heightened awareness of their individual environmental impact by paying attention to every item they purchased, used, or discarded.

Students also worked on creating greener solutions to every day design and communication problems, such as a campus-wide environmental awareness campaign and a re-design exercise. Using Re-nourish's Project Calculator and Paper Finder, students analyzed a found design then transformed it into a greener piece.

Discussions on environmental terminology, certifications, and eco-labels allowed students to recognize truly greener products and organizations from those that greenwash.

Guest lecturers, such as Re-nourish's own Jess Sand, shared important real-world knowledge about being an environmentally responsible designer today, running a responsible design studio, and talking to clients about sustainability. Jess' lecture was helpful in linking theory to practice.

SUSTAINABILITY ASSESSMENT (SA)

Sustainability Assessment is an umbrella term that can include a range of approaches or methodologies such as Sustainability Appraisal, Sustainability Impact Assessment, Integrated Sustainability Assessment5, or Integrated Assessment6, amongst others. Sustainability Assessment may be conducted by regulators for approval purposes, in what might be termed "external" Sustainability Assessment; while "internal" Sustainability Assessment is increasingly conducted by proponents themselves as a tool to improve internal decision-making and the overall sustainability of the final proposal. It is being applied to an ever-increasing range of decisions across the world, from policies, to strategic plans, to projects to trade agreements, at different levels (local, regional, national, international, sectoral) and with different timing (ex ante, during, ex post).

There are many variations of the forms of sustainability-based assessment existing throughout the world. Many international organisations and associations have adopted sustainability and have been developing indicators for sustainability assessments. Some of these organisations and associations include Organisation for Economic Cooperation and Development (OECD), World Trade Organisation (WTO), World Economic Forum's (Pilot Environmental Sustainability Index), the Dow Jones Sustainability Indexes, the Global Mining Initiative (GMI) Mining Minerals and Sustainable Development (MMSD) assessment process (assessing the state of the mining industry in terms of sustainability) and European Commission's Sustainable Development Indicators (SDIs). Some of the other models which are being used by industries include the Sustainability Assessment Model (SAM) – used by BP and SPeAR (Arup's spreadsheet and diagram model).

More recently, the EU has introduced the more comprehensive tool of (Sustainability) Impact Assessment. The intention is to move from the sectoral and often fragmented assessments to an integrated assessment covering environmental, economic and social parameters. The goal of this new tool is to be able to identify the likely positive and negative impacts of proposed policy actions – notably those relevant under the EU Sustainable Development Strategy – and thus enable informed political judgements about the proposal. Furthermore, as part of FP6, three other research projects: SustainabilityA-Test, MATISSE, and FORESCENE, have investigated options and developed analytical frameworks for Sustainability Assessment.

Although sustainability is a multi-dimensional integrative concept, it is context-bound and needs to be interpreted and implemented by a range of stakeholders within that specific context. The concept of sustainable development is contested, both scientifically and socially; therefore Sustainability Assessment could also be subjective and ambiguous. The SustainabilityA-test project suggests that sustainability assessments should bring together as many relevant aspects in the context of sustainable development as possible (SutainabilityAtest website). These include human and biophysical, present and future, local and global, active and precautionary, critique and alternative vision, concept and practice, and universal and context-specific. In addition, proper sustainability implementation should engage together participants covering the full range of public, corporate and civil society organisations and institutions, as well as individuals with their various capacities and inclinations. And all of these are recognised as constituent factors in complex and dynamic interrelations. Appreciation of uncertainty is also part of the sustainability concept. Still, the essence of the concept, and the key to its implementation, is clearly centred on appreciation of links and integration of the relevant considerations.

Often in sustainability assessments, the three pillars (social, economic and ecological) are used as conventional disciplinary categories to represent the main broad areas of concerns as well as for the structuring of sustainability indicators. Box 1 presents examples of different impacts to be evaluated under the EU Impact Assessment for these three categories. Although this approach is comprehensive and familiar way of organising sustainability assessment criteria, it does not integrate and deal with cross – pillar issues. In this respect, Gibson *et al.* (2005) suggest the following generic assessment criteria for the sustainability assessment framework. The authors argue that these criteria not only cover all core sustainability requirements, but also force thinking across the boundaries between the three usual pillar categories, and draw explicit attention to the concerns most commonly ignored or marginalised in conventional decision-making.

Under the SA framework, various tools can be used to assess these impacts. However, a common problem identified in literature is the lack of guidance on what tools can be used. Frequently, guideline documents provide the necessary procedural steps, checklists and matrices but remarkably little actual methodological and analytical guidance. Literature and some research projects recommend that sustainability assessment process should combine various existing assessment tools and indicators to help decision-making. For example, Azapagic (2003) and Azapagic and Perdan (2000) propose a life cycle approach to sustainability assessment of industrial systems, using LCA and indicators of sustainable development as the tools. Sustainability A-Test realises that assessing sustainability involves "multiple generations (*i.e.* longer time scales), multiple geographical scales (*i.e.* from local to global), multiple domains (*i.e.* economic, environmental and social) and multiple perspectives (*i.e.* different ideas about how to develop sustainable)". To include such wide aspects, the project maps several existing tools which could be used in combination. These include:

1. Physical assessment tools

2. Monetary assessment tools

3. Models (various computer models)

4. Scenario analysis

5. Multi-criteria analysis

6. Sustainability/environmental appraisal tools

7. Participatory tools (aiming to involve stakeholders)

8. Transition management

Strengths

SA provides a flexible framework for evaluation of different systems at all levels, from micro to macro, with the main goal to merge different approaches and methodologies into overarching sustainability assessments and to identify synergies and trade-offs among the different sustainability dimensions.

The framework allows using combination of various tools to assess different sustainability (economic, environmental and social) aspects.

It links the human and biophysical, present and future, local and global, active and precautionary, critique and alternative vision, concept and practice, as well as universal and context-specific aspects.

Weaknesses/Limitations

The research on how to organise and deploy different tools and methodologies in assessments is still in its infancy. A common problem identified in literature is the lack of guidance on what tools can be used. Often, guideline documents provide the necessary procedural steps, checklists and matrices but little actual methodological and analytical guidance.

Often many approaches to sustainability oriented assessments – at the project as well as strategic level – address the social, economic and ecological considerations separately and then struggle with how to integrate the separate findings.

Integration of qualitative and quantitative information into a single framework is also a critical issue for Sustainability Assessment.

Opportunities for Broadening And Deepening LCA

SA facilitates broader sustainability policy integration in every political or strategic decision.

In SA there is a strong emphasis on stakeholder engagement as well as inducing a reframing and learning process among participants in the process.

In 2002, the EU introduced (Sustainability) Impact Assessment as a comprehensive tool to identify the likely positive and negative impacts of proposed policy actions – notably those relevant under the EU Sustainable Development Strategy – and thus enable informed political judgements about the proposal.

Threats for Broadening and Deepening LCA

The concept of sustainable development is contested, both scientifically and socially, therefore Sustainability Assessment could also be subjective, ambiguous, context-bound and needs to be interpreted and implemented by a range of stakeholders within that specific context.

Integrating environmental, economic and social aspects is not easy and may deter end-users from engaging in the process.

The data requirements are substantial and may lead to protracted analyses, discussions or abandoning the assessment.

PROBLEM SOLVING FOR SUSTAINABILITY

It should be clear by now that making decisions and solving problems in support of greater sustainability of human-created systems and their impact on the natural environment is a complex undertaking. Often in modern life our decisions and designs are driven by a single goal or objective (*e.g.* greater monetary profitability, use of less energy, design for shorter travel times, generation of less waste, or reduction of risk), but in most cases solving problems sustainably requires a more holistic approach in which the functioning of many parts of the system must be assessed simultaneously, and multiple objectives must be integrated when possible. Furthermore, as noted in the Brundtland Report, often our decisions require the recognition of tradeoffs – there are many kinds of impacts on the environment and most decisions that we make create more than one impact at the same time. Of course choices must be made, but it is better if they are made with fuller knowledge of the array of impacts that will occur. The history of environmental degradation is littered with decisions and solutions that resulted in unintended consequences.

An illustrative example of the role of sustainability in solving problems is the issue of biofuels – turning plant matter into usable energy (mostly liquid hydrocarbon-based fuels). When viewed from afar and with a single goal, "energy independence," using our considerable agricultural resources to turn solar energy, via photosynthesis, into usable fuels so that we can reduce our dependence on imported petroleum appears to be quite attractive. The United States is the largest producer of grain and forest products in the world. It has pioneered new technologies to maintain and even increase agricultural productivity, and it has vast processing capabilities to create artificial fertilizer and to convert biomass into agricultural products.

And, after all, such a venture is both "domestic" and "natural" – attributes that incline many, initially at least, to be favorably disposed. However upon closer examination this direction is not quite as unequivocally positive as we might have thought. Yes it is possible to convert grain into ethanol and plant oils into diesel fuel, but the great majority of these resources have historically been used to feed Americans and the animals that they consume (and not just Americans; the United States is the world's largest exporter of agricultural products). As demand has increased, the prices for many agricultural products have risen, meaning that some fraction of the world's poor can no longer afford as much food. More marginal lands (which are better used for other crops, grazing, or other uses) have been brought under cultivation for fermentable grains, and there have been parallel "indirect" consequences globally – as the world price of agricultural commodities has risen, other countries have begun diverting land from existing uses to crops as well. Furthermore, agricultural runoff from artificial fertilizers has contributed to over 400 regional episodes of **hypoxia** in estuaries around the world, including the U.S. Gulf Coast and Chesapeake Bay.

In response to such problems, U.S. Congress passed the Energy Independence and Security Act in 2007, which limits the amount of grain that can be into biofuels in favor of using agriculturally-derived cellulose, the chief constituent of the cell walls of plants. This has given rise to a large scientific and technological research and development program to devise economical ways to process cellulosic materials into ethanol, and parallel efforts to investigate new cellulosic cropping systems that include, for example,native grasses. Thus, the seemingly simple decision to grow our biofuels industry in response to a political objective has had unintended political, financial, dietary, social, land use, environmental quality, and technological consequences.

With hindsight, the multiple impacts of biofuels have become clear, and there is always the hope that we can learn from examples like this. But we might also ask if there is a way to foresee all or at least some of these impacts in advance, and adjust our designs, processes, and policies to take them into account and make more informed decisions, not just for biofuels but also for complex societal problems of a similar nature. This approach is the realm of the field of **industrial ecology**, and the basis for the tool of **life cycle assessment (LCA)**, a methodology that has been designed to perform holistic analyses of complex systems.

Industrial Ecology

Many systems designed by humans focus on maximizing profitability for the firm, business or corporation. In most cases this means increasing production to meet demand for the products or services being delivered. An unfortunate byproduct of this is the creation of large amounts of waste, many of which have significant impacts if they enter the environment. A general-purpose diagram of a typical manufacturing process, showing the inputs of materials and energy, the manufacturing of products, and the generation of wastes (the contents of the "manufacturing box" are generic and not meant to depict any particular industry—it could be a mine, a factory, a power plant, a city, or even a university). What many find surprising is the large disparity between the amounts of waste produced and the quantity of product delivered. Table **Waste-to-Product Ratios for Selected Industries** provides such information, in the form of waste-to-product ratios, for a few common industries.

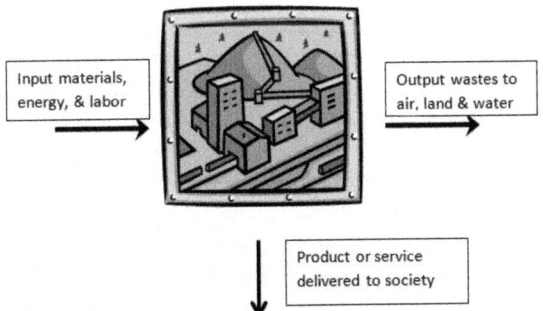

Human-Designed Industry Generic representation of a human-designed industry.

That industrial systems designed to maximize production and/or profits while ignoring wastes should be so materially inefficient is not surprising. As noted in the Module **Sustainability and Public Policy**, the impacts of wastes on human health and the environment have historically been ignored, or steeply underpriced, so that little incentive has existed to limit waste production. More recently laws have been enacted that attempt to force those responsible for waste emissions into a more appropriate accounting. Once realistic costs are assigned to the waste sector, manufacturers are quick to innovate and investigate ways to eliminate them.

Waste-to-Product Ratios for Selected Industries: Table shows the waste to product ratios for six common industries.

Industrial Sector	Waste-to-Product Ratio
Automobiles	2/1 (up to 10/1 if consumer use is included)
Paper	10/1
Basic Metals (*e.g.* Steel and Aluminum)	30-50/1
Chemicals	0.1-100/1
Nanostructured materials (*e.g.* computer chips)	700-1700/1
Modern Agriculture	~4/1

In 1989, Robert Frosch & Nicholas Gallopoulos, who worked in the GENERAL MOTORS Research Laboratory, published an important analysis of this problem in Scientific American. Their paper was entitled "Strategies for Manufacturing"; in it they posed a critical question: Why is it that human-designed manufacturing systems are so wasteful, but systems in nature produce little, if any, waste? Although there had been many studies on ways to minimize or prevent wastes, this was the first to seek a systemic understanding of what was fundamentally different about human systems in distinction to natural systems. The paper is widely credited with spawning the new field of Industrial Ecology, an applied science that studies material and energy flows through industrial systems. Industrial Ecology is concerned with such things as closing material loops (recycling and reuse), process and energy efficiency, organizational behavior, system costs, and social impacts of goods and services. A principle tool of Industrial Ecology is life cycle assessment.

Life Cycle Assessment Basics

LCA is a systems methodology for compiling and evaluating information on materials and energy as they flow through a product or service manufacturing chain. It grew out of the needs of industry, in the early 1960s, to understand manufacturing systems, supply chains, and market behavior, and make choices among competing designs, processes, and products. It was also applied to the evaluation of the generation and emission of wastes from manufacturing activities.

During the 1970s and 1980s general interest in LCA for environmental evaluation declined as the nation focused on the control of toxic substances and remediation of hazardous waste sites, but increasing concern about global impacts, particularly those associated with greenhouse gas emissions, saw renewed interest in the development of the LCA methodology and more widespread applications.

LCA is a good way to understand the totality of the environmental impacts and benefits of a product or service. The method enables researchers and practitioners to see where along the product chain material and energy are most intensively consumed and waste produced. It allows for comparisons with conventional products that may be displaced in commerce by new products, and helps to identify economic and environmental tradeoffs.

LCA can facilitate COMMUNICATION of risks and benefits to stakeholders and consumers (*e.g.* the "carbon footprint" of individual activities and life styles). Perhaps most importantly of all, LCA can help to prevent unintended consequences, such as creating solutions to problems that result in the transferal of environmental burdens from one area to another, or from one type of impact to another.

A complete LCA assessment defines a system as consisting of four general stages of the product or service chain, each of which can be further broken down into substages:

- Acquisition of materials (through resource extraction or recycled sources)
- Manufacturing, refining, and fabrication
- Use by consumers
- End-of-life disposition (incineration, landfilling, composting, recycling/reuse)

Each of these involves the transport of materials within or between stages, and transportation has its own set of impacts.

In most cases, the impacts contributed from each stage of the LCA are uneven, *i.e.* one or two of the stages may dominate the assessment. For example, in the manufacture of aluminum products it is acquisition of materials (mining), purification of the ore, and chemical reduction of the aluminum into metal that create environmental impacts. Subsequent usage of aluminum products by consumers contributes very few impacts, although the facilitation of recycling of aluminum is an important step in avoiding the consumption of primary materials and energy. In contrast, for internal combustion-powered AUTOMOBILES, usage by consumers creates 70-80% of the life cycle impacts. Thus, it is not always necessary that the LCA include all stages of analysis; in many cases it is only a portion of the product/service chain that is of interest, and often there is not enough information to include all stages anyway. For this reason there are certain characteristic terminologies for various "scopes" of LCAs that have emerged:

- *Cradle-to-grave*: includes the entire material/energy cycle of the product/material, but excludes recycling/reuse.

- *Cradle-to-cradle*: includes the entire material cycle, including recycling/reuse.

- *Cradle-to-gate*: includes material acquisition, manufacturing/refining/fabrication (factory gate), but excludes product uses and end-of-life.

- *Gate-to-gate*: a partial LCA looking at a single added process or material in the product chain.

- *Well-to-wheel*: a special type of LCA involving the application of fuel cycles to transportationvehicles.

- *Embodied energy*: A cradle-to-gate analysis of the life cycle energy of a product, inclusive of the latent energy in the materials, the energy used during material acquisition, and the energy used in manufacturing intermediate and final products. Embodied energy is sometimes referred to as "emergy", or the cumulative energy demand (CED) of a product or service.

LCA Methodology

Over time the methodology for conducting Life Cycle Analyses (LCAs) has been refined and standardized; it is generally described as taking place in four steps: scoping, inventory, impact assessment, and interpretation. The first three of these are consecutive, while the interpretation step is an ongoing process that takes place throughout the methodology.

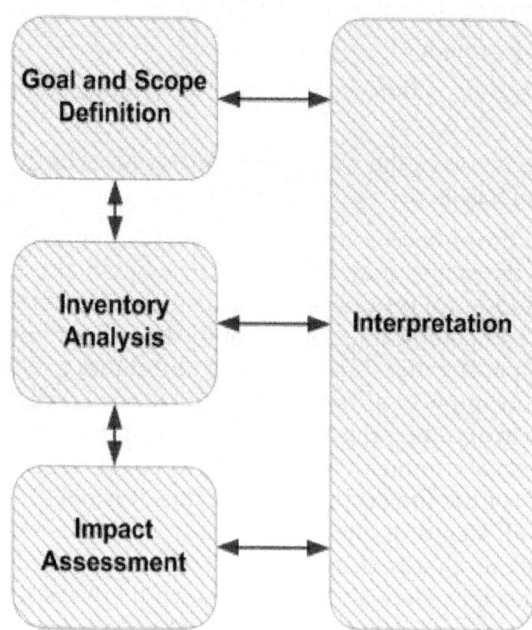

Fig. : General Framework for Life Cycle Assessment.

The four steps of life cycle assessment and their relationship to one another.

Scoping

Scoping is arguably the most important step for conducting an LCA. It is here that the rationale for carrying out the assessment is made explicit, where the boundaries of the system are defined, where the data quantity, quality, and sources are specified, and where any assumptions that underlie the LCA are stated. This is critically important both for the quality of the resultant analysis, and for comparison among LCAs for competing or alternative products.

Inventory Analysis

The inventory analysis step involves the collection of information on the use of energy and various materials used to make a product or service at each part of the manufacturing process. If it is true that scoping is the most important step in an LCA then the inventory is probably the most tedious since it involves locating, acquiring, and evaluating the quality of data and specifying the sources of uncertainties that may have arisen. For products that have been produced for a long time and for which manufacturing processes are well known, such as making steel, concrete, paper, most plastics, and many machines, data are readily available. But for newer products that are either under development or under patent protection, data are often considered proprietary and are generally not shared in open sources.

Uncertainty can arise because of missing or poorly DOCUMENTED data, errors in measurement, or natural variations caused by external factors (*e.g.*, weather patterns can cause considerable variation in the outputs of agricultural systems or the ways that consumers use products and services can cause variability in the emission of pollutants and the disposition of the product at end of life). Often the manufacturing chain of a process involves many steps resulting in a detailed inventory analysis. The manufacturing flow for a bar of soap (this diagram is for making bar soap using saponification — the hydrolysis of triglycerides using animal fats and lye). The inventory requires material and energy inputs and outputs for each of these steps, although it may turn out that some steps contribute little to the ultimate impact analysis. For example, the inventory associated with capital equipment for a manufacturing process, *i.e.* machines that are replaced at lengthy intervals such that their impacts in the short term are minimal, are often omitted from the analysis.

There are two additional aspects of LCA that should also be addressed during inventory analysis: the **functional unit** of comparison, and the **allocation** of inventory quantities among co-products or services. The functional unit is the basis for comparing two or more products, processes, or services that assure equality of the function delivered. This may seem like a straightforward task. For example, for the soap produced by the process, one might choose "one bar of soap" as a functional unit of comparison. But then how would a LCA comparison be made with, say, liquid hand soap or a body wash product (which combines the functionality of soap and shampoo)? Perhaps "number of washings" would be a better choice, or maybe concentration of surfactant made available per average use (in the latter case an "average dose" would need to be defined). Furthermore, soaps have other additives and attributes such as scents, lotions, colors, and even the functionality of the shape – factors that may not affect cleaning effectiveness but certainly do have an impact on consumer preferences, and hence quantity sold. Since it is quite likely that essentially all soaps purchased by consumers will eventually be washed down the drain, such marketability factors may indeed have an environmental impact.

Inventory data are virtually always sought for a total supply-manufacturing-consumer-use chain rather than individual products, thus when that same chain produces multiple products it is necessary to allocate the materials, energy, and wastes among them. Again,, there are potentially several co-products produced: tallow and other animal products, forest products, cardboard and paper, and salable scrap. There are generally three ways to allocate materials and energy among co-products: mass, volume, and economic value. Mass and volume allocations are the most straightforward, but may not capture market forces that are important in bringing materials into the environment. Allocation via economic valuation usually reflects the value of the energy and any "value added" to the raw materials, but may miss the impacts of the materials themselves. In addition, market values may fluctuate over time. In the final analysis the important aspect of any allocation procedure is that it be fully documented.

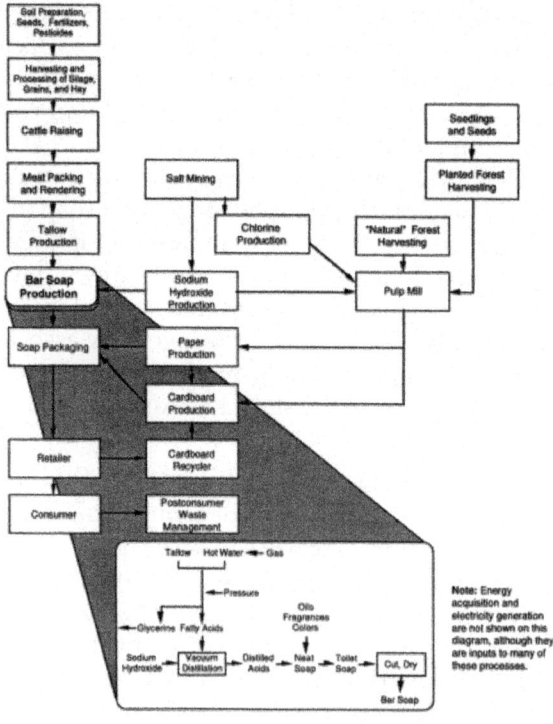

Fig. : Detailed System Flow Diagram for Bar Soap.

The manufacturing flow for a bar of soap (this diagram is for making bar soap using saponification – the hydrolysis of triglycerides using animal fats and lye).

Impact Assessment

The life cycle impact assessment (LCIA) takes the inventory data on material resources used, energy consumed, and wastes emitted by the system and estimates potential impacts on the environment. At first glance, given that an inventory may include thousands of substances, it may seem that the number of potential impacts is bewilderingly large, but the problem is made more tractable through the application of a system of impact classifications within which various inventory quantities can be grouped as having similar consequences on human health or the environment. Sometimes inventoried quantities in a common impact category originate in different parts of the life cycle and often possess very different chemical/biological/physical characteristics. The LCIA groups emissions based on their common impacts rather than on their chemical or physical properties, choosing a reference material for which health impacts are well known, as a basic unit of comparison. A key aspect is the conversion of impacts of various substances into the reference unit. This is done using characterization factors, some of which are well-known, such as global warming potential and ozone depletion potential, and LC_{50} (the concentration of a substance at which fifty percent of an exposed population is killed), and others are still under development. Table **Common Impact Categories and Their References** presents several impact categories that are frequently used in the LCIA along with their references. The categories listed in Table **Common Impact Categories and Their References** are not exhaustive – new types of impact categories, such as land use and social impacts – and continue to be developed.

Common Impact Categories and Their References: Several impact categories that are frequently used in the LCIA along with their references.

Human Health (cancer)	Kg Benzene eq/unit
Human Health (non-cancer)	LC_{50} eq from exposure modeling
Global Climate Change	Kg CO_2 eq/unit
Eutrophication	Kg Nitrogen eq/unit
Ecotoxicity Aquatic, Terrestrial Toxicity	Kg 2,4 D eq/unitLC$_{50}$ eq from exposure modeling
Acidification	Kg H^+/unit
Smog Formation	Kg Ethane eq/unit
Stratospheric Ozone Depletion	Kg CFC-11 eq/unit

An example will help to illustrate the type of information that results from life cycle inventory and impact assessments. In this case, a system that produces a biologically-derived plastic, polylactide, is examined. (PLA). PLA has been proposed as a more sustainable alternative to plastics produced from petroleum because it is made from plant materials, in this case corn, yet has properties that are similar to plastics made from petroleum. A schematic of the system, which is a cradle-to-gate assessment. As with any plastic, PLA can be turned into a variety of final products and each will have different cradle-to-grave LCA characteristics. The production of PLA involves growing corn, harvesting and processing the grain, and polymerizing the lactic acid molecules produced from fermentation. At each step a variety of chemicals and energy are used or produced. It is these production materials that contribute to the impact analysis. Inventory quantities were allocated among major bio-products on a mass basis.

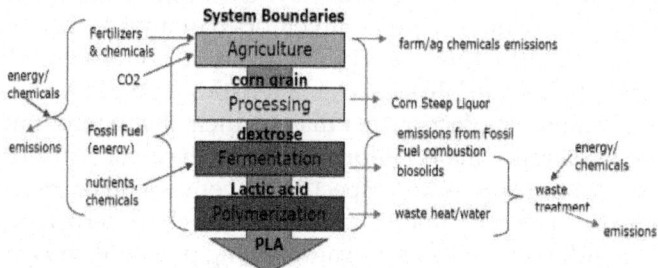

Fig. : Processing Diagram for Making Polylactide (PLA) .

The production of PLA involves growing corn, harvesting and processing the grain, and polymerizing the lactic acid molecules produced from fermentation. At each step a variety of chemicals and energy are used or produced. It is these production materials that contribute to the impact analysis.

Among the inventory data acquired in this case is life cycle fossil fuel used by the system, mostly to power farming equipment ("Agriculture"), wet-mill corn ("CWM"), heat fermentation vats ("Ferment"), and Polymerization ("Polym"). The transport of intermediate products from sources to the processing center is also included. The fossil fuel used to make PLA compared with fossil fuel used for making several petroleum-based plastics. The global warming potential impact analysis.

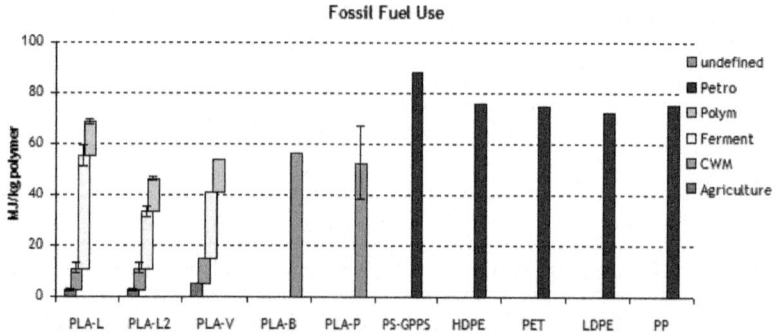

Fig. : Fossil Fuel Use to Make PLA *vs.* Petroleum-Based Plastics.

The amount of fossil fuels used when making PLA is slightly less in comparison to making several petroleum-based products.

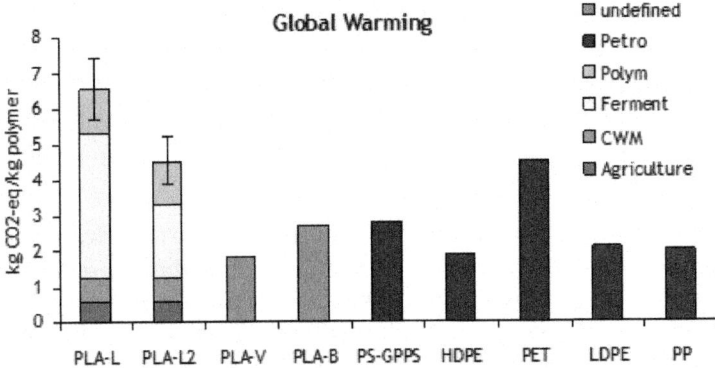

Fig. : Global Warming Potential Impact Analysis.

Global warming impact for PLA compared with several other petroleum-derived plastics.

Equal mass basis (the functional unit is one kilogram of plastic). The PLA inventory also shows the sources of fossil fuel used for each step along the manufacturing chain, with the fermentation step being the most intense user. What may not be obvious is that the total greenhouse gases (GHG) emitted from the process, on an equivalent carbon dioxide (CO_2) basis, are generally higher for the biopolymer in comparison with the petroleum polymers in spite of the lower fossil fuel usage. When the data are examined closely this is due to the agricultural step, which consumes generates relatively little fossil fuel, but is responsible for a disproportionate amount of emissions of GHGs, mostly in the form of nitrous oxide, a powerful greenhouse gas (310 times the global warming potential of CO_2) that is a by-product of fertilizer application to fields. This example also illustrates counter-intuitive results that LCAs often generate, a principal reason why it is important to conduct them.

Interpretation of LCA

The interpretation step of LCA occurs throughout the analysis. Issues related to the rationale for conducting the LCA, defining the system and setting its boundaries, identifying data needs, sources, and quality, and choosing functional units, allocation procedures, and appropriate impact categories must all be addressed as the LCA unfolds. There are essentially two formal reasons for conducting an LCA: (a) identification of "hot spots" where material and/or energy use and waste emissions, both quantity and type, are greatest so that efforts can be focused on improving the product chain; and (b) comparison of results between and among other LCAs in order to gain insight into the preferable product, service, process, or pathway. In both cases, there are cautions that apply to the interpretation of results.

Assumptions

Typically a variety of assumption must be made in order to carry out the LCA. Sometimes these are minor, for example, exclusion of elements of the study that clearly have no appreciable impact on the results, and sometimes more critical, for example choosing one set of system boundaries over another. These must be explicitly stated, and final results should be interpreted in light of assumptions made

Data Quality, Uncertainty, and Sensitivity

In the course of conducting an LCA it is usually the case that a variety of data sources will be used. In some cases these may be from the full-scale operation of a process, in others the source is from a small scale or even laboratory scale, in still other cases it may be necessary to simulate information from literature sources. Such heterogeneity inevitably leads to uncertainty in the final results; there are several statistical methods that can be applied to take these into account. An important aspect of the completed LCA is the degree of sensitivity the results display when key variables are perturbed. Highly sensitive steps in the chain have a greater need to narrow uncertainties before drawing conclusions with confidence.

Incommensurability

Sometimes LCA impact categories, such as those shown in Table **Common Impact Categories and Their References**, overlap in the sense that the same pollutant may contribute to more than one category. For instance, if a given assessment comes up with high scores for both aquatic toxicity and human toxicity from, say, pesticide use then one might be justified in using both of these categories to draw conclusions and make choices based on LCA results. However, more typically elevated scores are found for categories that are not directly comparable. For instance, the extraction, refining, and use of petroleum generate a high score for global warming (due to GHG release), while the product chain for the biofuel ethanol has a high score for eutrophication (due to nitrogen release during the farming stage). Which problem is worse – climate change or coastal hypoxia? Society may well choose a course of action that favors one direction over another, but in this case the main value of the LCA is to identify the tradeoffs and inform us of the consequences, not tell us which course is "correct."

Risk Evaluation and Regulation

One of the inherent limits to LCA is its use for assessing risk. Risk assessment and management, as described in the Modules The Evolution of Environmental Policy in the United States and Modern Environmental Management, is a formal process that quantifies risks for a known population in a specific location exposed to a specific chemical for a defined period of time. It generates risk values in terms of the probability of a known consequence due to a sequence of events that are directly comparable, and upon which decisions on water, land, and air quality standards and their violation can be and are made. LCA is a method for evaluating the impacts of wastes on human health and the environment from the point of view of the product/service chain rather than a particular population. It can be used to identify the sources of contamination and general impacts on the environment – a sort of "where to look" guide for regulation, but its direct use in the environmental regulatory process has been, to date, rather limited. One application for LCA that has been suggested for regulatory use is for assessing the impacts of biofuel mandates on land use practices, in the United States and other regions, however no regulatory standards for land use have yet been proposed.

THERMODYNAMIC FOUNDATIONS

Thermodynamics studies energy transformations and entropy changes across the entire realm of micro- and macroscopic processes. Entropy's probabilistic measure, given by statistical mechanics, turns out to be equally useful for measuring information and complexity, revealing the affinity between the three. As such, thermodynamic laws are independent of the system's specific chemistry or types of energies involved. This ubiquity makes thermodynamics a powerful framework for studying all organic processes, even where the system's specific details are not accessible.

We begin with introducing the concepts of energy, work and entropy. The laws of thermodynamics are then explained. Energy's various types and transformations are presented, and energy conservation, dictated by the First Law, is demonstrated. Next, illustrated by Carnot's cycle, comes the non-conserved quantity, namely, entropy. The Second Law is thus introduced. Entropy is studied by all its manifestations and measures: probability, disorder, irreversibility, heat, *etc.* Students are urged to explore their differences and overlaps until comprehending all of them, both intuitively and mathematically. We next proceed to the concepts of information and complexity, shown to be hallmarks of all living systems. With the aid of "Maxwell's Demon" paradox, their thermodynamic measures are shown to follow from those of entropy.

The energy-information tradeoff thus serves as an introduction to the course's second part. Basic biological notions are introduced from evolutionary theory, population genetics, nanotechnology and environmental sciences, showing how their entropy, information content and complexity can be measured with predictive power for other biological measures. Critical discussion of open questions, fallacies and misuses of thermodynamics are added. We conclude with pointing out further possible applications of thermodynamics to life sciences.

Free Energy and Composition

To lay this foundation, the tool that will prove to be most useful is the concept of **equilibrium**. Equilibrium is best described in terms of **free energy**. Free energy is a measure of a systems internal energy which gives an indication of the randomness or **entropy** of the system. The best way to use free energy to describe a phase diagram is to use a graph of free energy versus the composition (a G-X diagram) of a system.

For any given phase a G-X plot can be made holding temperature and pressure constant.

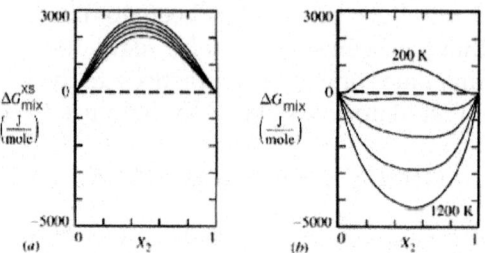

Fig. : Free energy-compostion diagrams for a flexible three parameter solution model plotted as a function of temperature.

Each of these curves is represented by the mathematical equation:

$$\Delta G_{min} = RT\,(X_1\ln X_1 + X_2\ln X_2)$$

The equation for the G-X curve of a real solution is given by:

$$\Delta G_{min} = \Delta G_{min}^{XS} + RT(X_1\ln X_1 + X_2\ln X_2)$$

where the excess free energy of mixing generally depends on composition, temperature, and pressure. The excess free energy contributes to ideal free energy of mixing. Now that each phase that may exist in its system has its own G-X curve, phases may be compared. However, in order to compare G-X curves of different phases, the energies of each component in the system must be related to the same reference state.

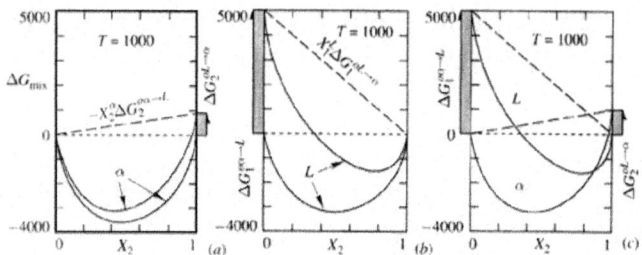

Fig. : Effect of change in refernece states of pure 1 and pure 2 upon the shape of the G-X diagram for (a) the solid solution, alpha, and (b) the liquid solution L. The comparison of mixing behavior in (c) is now valid because the reference states are consistent.

Reference States

G-X curves that are constructed for each phase are derived from the mixing processes in which the solution was formed from pure components in some initial (and unmixed) condition. The **reference state** is this unmixed condition.

Since there is a G-X curve for each phase, there must be a reference state for each component in the system. Specification of any reference state requires that each component of the phase diagram at the beginning of the mixing process has four conditions:

- Pressure in the component = pressure in the solution
- Temperature in component = temperature in the solution
- Composition in component = that of a pure component
- Phase Form in component = phase form in the solution.

So, for example, if one needs to describe an ideal liquid solution with components X and Y at a standard temperature and pressure, the reference state for X is pure liquid X at STP; the reference state for Y is pure liquid Y at STP. Similarly, in an ideal solid solution of X and Y, the reference states are pure solid X and Y respectively at STP. However, to relate solid solutions with liquid solutions to compare energies of mixing, these models are not enough.

To compare liquid and solid solutions, one must find a reference state that is the same for both of the solutions. To do this, it is required to change the reference states in the mixing calculations of one of the solutions. To change one of the reference states requires a knowledge of Gibbs free energies between the different reference states.

Changing equation 1 in such a way as to make its reference state explicitly known:

$$\Delta G_{min}^a (a; a) = X_X^a (\bar{G}_X^a - G_X^{oa}) + X_Y^a (\bar{G}_Y^a - G_Y^{oa})$$
$$\Delta G_{min}^L (L; L) = X_X^L (\bar{G}_X^L - G_X^{oL}) + X_Y^L (\bar{G}_Y^L - G_Y^{oL})$$

where a and L represent the phases of the X and Y components the notation in the brackets represent the reference states for the X and Y components. $\bar{G}$, G^o are Gibbs free energies of any component in solution and in its pure form respectively.

Now, to compare the mixing behavior of the two solutions the reference states need to be the same. There is more than one way to accomplish this as one can see in this table:

	I	II	III	IV
Ref. state for alpha solution	{alpha;alpha}	{alpha;L}	{L;alpha}	{L;L}
		or	or	or
Ref. state for liquid solution	{alpha;alpha}	{alpha;L}	{L;alpha}	{L;L}

To illustrate how to change reference states, let us choose reference state labeled II. Looking at this state, we have to calculate the following equations:

$$\Delta G_{min}^a (a; L) = X_X^a (\bar{G}_X^a - G_X^{oa}) + X_Y^a (\bar{G}_Y^a - G_Y^{oL})$$
$$\Delta G_{min}^L (a; L) = X_X^L (\bar{G}_X^L - G_X^{oa}) + X_Y^L (\bar{G}_Y^L - G_Y^{oL})$$

Notice that the G^{oa}_Y term in the given equation has been replaced by G^{oL}_Y in the given equation and G^{oL}_X in the given equation has been replaced by G^{oa}_X in the given equation - this is done as specified by the reference states on the left side of the equality. The G^o terms stay the same for the X and Y components which represent the a and L phases respectively. Now the energies of mixing can be compared.

To get these new solution models in terms of some explicit reference state, we need to reference them to in the given equations with their assumed reference states. When we do this, we get:

$$\Delta G^a_{min}(a;L) = X^a_X(\overline{G}^a_X - G^{oa}_X) + X_Y(\overline{G}^a_Y - G^{oa}_Y) + X^a_X(G^{oa}_Y - G^{oL}_Y)$$
$$\Delta G^L_{min}(a;L) = X^L_X(\overline{G}^L_X - G^{oL}_X) + X_Y(\overline{G}^L_Y - G^{oL}_Y) + X^L_X(G^{oL}_X - G^{oa}_X)$$

As is seen, the first two terms in the models are identical to in the given equations. For simplicity sake we refer to the last term as:

$$G^{oa}_Y - G^{oL}_Y = \Delta G^{oL \to a}_Y$$
$$G^{oL}_X - G^{oa}_X = \Delta G^{oa \to L}_X$$

These equations could be thought of as being related to the Gibbs free energy of melting, moving from one phase to another.

Finally, the Gibbs free energy of mixing can be expressed as:

$$\Delta G^a_{min}(a;L) = \Delta G^a_{min}(a;a) + X^a_Y \Delta G^{oL \to a}_Y$$
$$\Delta G^a_{min}(a;L) = \Delta G^L_{min}(L;L) + X^L_Y \Delta G^{oa \to L}_X$$

where the first terms can be computed from a solution model as in the given equation. The last terms require knowledge of Gibbs free energies between the solid and liquid phases from the pure components of the system.

This derivation was based on only one of four possible choices for equivalent reference states. Any other choice would lead to different linear terms. In general, the choice of reference states for the components determines the hanging points for the corresponding G-X curve.

Activity and Free Energy

A change in reference states also alters the calculated values of the **activity** ,a, and its coefficients for various components in that solution. This makes sense because activity is defined as:

$$\mu_k - \mu^o_k = RT \ln a_k$$

where mu is defined as the chemical potential of a component. If we apply this definition to component Y in its solid (alpha) phase, we get

$$\mu^a_Y - \mu^{oa}_Y = \overline{G}^a_Y - G^{oa}_Y = RT \ln a^a_Y$$

Keeping in mind that, at equilibrium, the change in free energy is zero as well as the change in chemical potential, mu. Thus,

$$\mu_{YL} - \mu_{YS} = 0,$$
$$\mu_{XL} - \mu_{XS} = 0$$

Applying this to in the given equation we get:

$$\mu_{YS} = \overline{G}_{YS} = G°_{YS} + RT \ln (a_{YS})$$
$$\mu_{XL} = \overline{G}_{XL} = G°_{XL} + RT \ln (a_{XL})$$
$$\mu_{YL} = \overline{G}_{YL} = G°_{YL} + RT \ln (a_{RL})$$
$$\mu_{XL} = \overline{G}_{XS} = G°_{XS} + RT \ln (a_{XS})$$

To keep the equations legible, terms such as $X_{Y,L}$ refer to the composition (X) of component Y in the (L)iquid phase.

While trying to keep this as simple as possible, from now on we will assume that we are dealing with an ideal solution. In an ideal solution, we shall also assume that the activity of a component at any phase is equal to its composition at that juncture.

$$a_{XL} = X_{XL}$$
$$a_{XS} = X_{XS}$$
$$a_{YS} = X_{YS}$$
$$a_{YL} = X_{YL}$$

Applying now last equation along with in the given equation; and keeping in mind that while the systems in the phase diagrams are in equilibrium, their free energy is at a minimum. In other words delta G_{mix} is zero. one can see:

$$G°_{YL} + RT \ln (X_{YL}) = G°_{YL} + RT \ln(X_{YS})$$
$$G°_{XL} + RT \ln (X_{XL}) = G°_{XS} + RT \ln(X_{XS})$$

In another effort to make the equations legible, Gibbs free energy of mixing from one phase to another, from its solid phase to liquid phase, a.k.a. free energy of melting will be denoted as delta G_M for any component of the system.

Using these ideal models and solving for the composition we get the equation of the **solidus line** for the Y component at any temperature, T, can be found and is given as follows

$$x_{YS}(T) = \frac{1 - \exp(-\Delta G°_{XM} / RT)}{\exp(-\Delta G°_{YM} / RT) - \exp(-\Delta G°_{XM} / RT)}$$

The equation for the **liquidus line** of component Y at any temperature, T, is found to be:

$$x_{YL}(T) = [x_{YS}(T)] [\exp(-\Delta G°_{YM} / RT)]$$

Keeping in mind that free energy of any substance can also be written as:

$$\Delta G = \Delta H - T \Delta S$$

where H is the enthalpy of a system and S is the familiar entropy. Enthalpy is a variable that takes into account the pressure and volume a system is at, an equation for gibbs free energy of any substance at some temperature, T, can be expressed as:

$$\Delta G^{a \to L} = \Delta H^{a \to L} \left(1 - \frac{T}{T^{a \to L}} \right).$$

STANDARD STATE

State of a system chosen as standard for reference by convention. Three standard states are recognized: For a gas phase it is the (hypothetical) state of the pure substance in the gaseous phase at the standard pressure $p = p°$, assuming ideal behaviour. For a pure phase, or a mixture, or a solvent in the liquid or solid state it is the state of the pure substance in the liquid or solid phase at the standard pressure $p = p°$. For a solute in solution it is the (hypothetical) state of solute at the standard molality $m°$, standard pressure $p°$ or standard concentration $c°$ and exhibiting infinitely dilute solution behaviour. For a pure substance the concept of standard state applies to the substance in a well defined state of aggregation at a well defined but arbitrarily chosen standard pressure.

In chemistry, the standard state of a material (pure substance, mixture or solution) is a reference point used to calculate its properties under different conditions. In principle, the choice of standard state is arbitrary, although the International Union of Pure and Applied Chemistry (IUPAC) recommends a conventional set of standard states for general use. IUPAC recommends using a standard pressure $p° = 10^5$ Pa. Strictly speaking, temperature is not part of the definition of a standard state. The standard state of a gas is conventionally chosen to be unit pressure (usually in bar) ideal gas, regardless of the temperature. However, most tables of thermodynamic quantities are compiled at specific temperatures, most commonly 298.15 K (25.00 °C; 77.00 °F) or, somewhat less commonly, 273.15 K (0.00 °C; 32.00 °F).

The standard state should not be confused with standard temperature and pressure (STP) for gases, nor with the standard solutions used in analytical chemistry.

At the time of development in the nineteenth century, the superscript plimsoll symbol was adopted to indicate the non-zero nature of the standard state. IU-PAC recommends in the 3rd edition of *Quantities, Units and Symbols in Physical Chemistry* a symbol which seems to be a degree sign (°) as a substitute for the plimsoll mark. In the very same publication the plimsoll mark appears to be constructed by combining a horizontal stroke with a degree sign. A range of similar symbols are used in the literature: a stroked lowercase letter O (⦵), a superscript zero (^{0}) or a circle with a horizontal bar either where the bar extends the boundaries of the circle or is enclosed by the circle, dividing the circle in half. When compared to the plimsoll symbol used on vessels, the horizontal bar should however extend the boundaries of the circle.

For a given material or substance, the standard state is the reference state for the material's thermodynamic state properties such as enthalpy, entropy, Gibbs free energy, and for many other material standards. The standard enthalpy change of formation for an element in its standard state is zero, and this convention allows a wide range of other thermodynamic quantities to be calculated and tabulated. The standard state of a substance does not have to exist in nature: for example, it is possible to calculate values forsteam at 298.15 K and 10^5 Pa, although steam does not exist (as a gas) under these conditions. The advantage of this practice is that tables of thermodynamic properties prepared in this way are self-consistent.

Conventional Standard States

Many standard states are non-physical states, often referred to as "hypothetical states". Nevertheless, their thermodynamic properties are well-defined, usually by an extrapolation from some limiting condition, such as zero pressure or zero concentration, to a specified condition (usually unit concentration or pressure) using an ideal extrapolating function, such as ideal solution or ideal gas behavior, or by empirical measurements.

Gases

The standard state for a gas is the hypothetical state it would have as a pure substance obeying the ideal gas equation at standard pressure (10^5 Pa, or 1 bar). No real gas has perfectly ideal behaviour, but this definition of the standard state allows corrections for non-ideality to be made consistently for all the different gases.

Liquids and Solids

The standard state for liquids and solids is simply the state of the pure substance subjected to a total pressure of 10^5 Pa. For most elements, the reference point of $\Delta H_f^\circ = 0$ is defined for the most stable allotrope of the element, such as graphite in the case of carbon, and the β-phase (white tin) in the case of tin. An exception is white phosphorus, the most common allotrope of phosphorus, which is defined as the standard state despite the fact that it is only metastable.

Solutes

For a substance in solution (solute), the standard state is the hypothetical state it would have at the standard state molality or amount concentration but exhibiting infinite-dilution behavior. The reason for this unusual definition is that the behavior of a solute at the limit of infinite dilution is described by equations which are very similar to the equations for ideal gases. Hence taking infinite-dilution behavior to be the standard state allows corrections for non-ideality to be made consistently for all the different solutes. Standard state molality is 1 mol kg^{-1}, while standard state amount concentration is 1 mol dm^{-3}.

The Standard States of the Elements

All chemical substances are either solid, liquid or gas. To make comparisons easier, the chemistry community has agreed on a concept called "the standard state." The standard state of a chemical substance is its phase (solid, liquid, gas) at 25.0 °C and one atmosphere pressure. This temperature/pressure combo is often called "room conditions."

Two elements are liquid in their standard state: mercury and bromine.

Eleven elements are gas in their standard state. All of the noble gases (He, Ne, Ar, Kr, Xe, Rn) as well the halogens flourine and chlorine. Hydrogen, oxygen and nitrogen are the others.

All other elements are solid in their standard state.

Some Interesting facts about elements:

1. Bromine's boiling point is 58.8 °C. Suppose you have a closed bottle of bromine liquid at 25 °C (room temperature). When you open the bottle, quite a bit of bromine vapor will be released. To get the idea, imagine a small puddle of water on the counter. Eventually, the puddle will all evaporate, even though the temperature never changed from 25 °C. In any event, when the ChemTeam was in high school chemistry (taken his junior year, 1968-69), he spied a bottle of bromine and wondered what it smelled like. Aha, the teacher is at the other end of the room, talking to someone. The ChemTeam grabbed the bottle, whipped the cap off and took a tenative sniff through the nose. Instant, major pain. Gagging. Great pain in the nose. Burning eyes. Did I mention the pain? Quickly, the cap was replaced and seat taken, with an innocent look plastered on the face. Man, that hurt!!

2. Iodine is a solid at room temperature, but it sublimes quite easily. This means it changes from solid directly to gas without going through the liquid state. Dry ice (solid carbon dioxide, which is not an element) does the same thing.

3). Mercury is a liquid AND a metal. That means it conducts electricity very easily. You might want to find out what a mercury switch is. The reasons why mercury is a liquid are complex, but evidently involve aspects of relativity (as in Albert Einstein).

4. Gallium is a solid at room conditions, but its melting point is just under 30 °C. In the past, a small amount of gallium was used to plug fire sprinklers, but I don't think gallium is used any more. Cesium has a melting point of 28.4 °C and I know for sure it is not used for fire sprinklers. Some atomic clocks (accurate to about one second in 3000 years) have used a "cesium fountain" as a timing device.

"DEAD STATE" IN THERMODYNAMICS

In 1865, Rudolf Clausius boldly and highly provocatively formulated that: *the energy of the universe remains constant and its entropy tends to a maximum*. Thus, because an increase in the entropy and disorder is associated with a decreased ability to perform work, the universe will ultimately reach a "dead state".

Thermodynamic State

For thermodynamics, a thermodynamic state of a system is fully identified by values of a suitable set of parameters known as state variables, state parameters or thermodynamic variables. Once such a set of values of thermodynamic variables has been specified for a system, the values of all thermodynamic properties of the system are uniquely determined.

Thermodynamics sets up an idealized formalism that can be summarized by a system of postulates of thermodynamics. Thermodynamic states are amongst the fundamental or primitive objects or notions of the formalism, in which their existence is formally postulated, rather than being derived or constructed from other concepts.

A thermodynamic system is not simply a physical system. Rather, in general, indefinitely many different alternative physical systems comprise a given thermodynamic system, because in general a physical system has vastly many more detailed characteristics than are mentioned in a thermodynamic description. A thermodynamic system is a macroscopic object, the microscopic details of which are not explicitly considered in its thermodynamic description. The number of state variables required to specify the thermodynamic state depends on the system, and is not always known in advance of experiment; it is usually found from experimental evidence. Always the number is two or more; usually it is not more than some dozen. Though the number of state variables is fixed by experiment, there remains choice of which of them to use for a particular convenient description; a given thermodynamic system may be alternatively identified by several different choices of the set of state variables.

For equilibrium thermodynamics, in a thermodynamic state of a system, its contents are in internal thermodynamic equilibrium, with zero flows of all quantities, both internal and between system and surroundings. For Planck, the primary characteristic of a thermodynamic state of a system that consists of a single phase, in the absence of an externally imposed force field, is spatial homogeneity. For non-equilibrium thermodynamics, a suitable set of identifying state variables includes some macroscopic variables, for example a non-zero spatial gradient of temperature, that indicate departure from thermodynamic equilibrium. Such non-equilibrium identifying state variables indicate that some non-zero flow may be occurring within the system or between system and surroundings.

State Functions

Besides the thermodynamic variables that originally identify a thermodynamic state of a system, the system is characterized by further quantities called state functions, which are also called state variables, thermodynamic variables, state quantities, or functions of state. They are uniquely determined by the thermodynamic state as it has been identified by the original state variables. A passage from a given initial thermodynamic state to a given final thermodynamic state of a thermodynamic system is known as athermodynamic process; it typically involve transfers of matter or energy between system and surroundings. In any thermodynamic process, whatever may be the intermediate conditions during the passage, the total respective change in the value of each thermodynamic state variable depends only on the initial and final states. For an idealized continuous process, this means that infinitesimal incremental changes in such variables are exact differentials. Together, the incremental changes throughout the process, and the initial and final states, fully determine the idealized process.

In the most commonly cited simple example, an ideal gas, the thermodynamic variables would be any two variables out of the following four: entropy, pressure, temperature, and volume. Thus the thermodynamic state would range over a two-dimensional state space. The remaining two variables, as well as other quantities such as the internal energy, would be expressed as state functions of these two variables. The state functions satisfy certain universal constraints, but ultimately they depend on the materials involved in the concrete system.

Various thermodynamic diagrams have been developed to model the transitions between thermodynamic states.

Equilibrium State

Physical systems found in nature are practically always dynamic and complex, but in many cases, macroscopic physical systems are amenable to description based on proximity to ideal conditions. One such ideal condition is that of a stable equilibrium state. Such a state is a primitive object of classical or equilibrium thermodynamics, in which it is called a thermodynamic state. Based on many observations, thermodynamics postulates that all systems that are isolated from the external environment will evolve so as to approach unique stable equilibrium states. There are a number of different types of equilibrium, corresponding to different physical variables, and a system reaches thermodynamic equilibrium when the conditions of all the relevant types of equilibrium are simultaneously satisfied. A few different types of equilibrium are listed below.

- Thermal Equilibrium: When the temperature throughout a system is uniform, the system is in thermal equilibrium.
- Mechanical Equilibrium: If at every point within a given system there is no change in pressure with time, and there is no movement of material, the system is in mechanical equilibrium.
- Phase Equilibrium: This occurs when the mass for each individual phase reaches a value that does not change with time.
- Chemical Equilibrium: In chemical equilibrium, the chemical composition of a system has settled and does not change with time.

BALANCES OF MASS, ENERGY, ENTROPY-DISSIPATION, AND AVAILABLE ENERGY

Energy and Entropy

Waste energy is associated with all processes. This waste can be reduced, but it can never be eliminated. Anyone who says otherwise is trying to con you.

The Second Law of Thermodynamics is a statement of the painfully obvious; that is, at least it's obvious to us now. The statements you are about to read were worked out over a period of many years, were refined and clarified by countless minds, and were gradually combined into one large idea that we now recognize as a comment on the fundamental behavior of the universe. Of all the relationships in physics given the title of "law", the Second Law of Thermodynamics is the one law for which there appears to be no exception. Anyone who disagrees with it or says they have a method to avoid it is immediately and rightfully dismissed as an idiot who doesn't understand the law, a crackpot pushing an agenda, or the perpetrator of some kind of hoax. When someone says they've found a way around the Second Law of Thermodynamics, chances are quite good they're trying to separate you from your money. Arguing against the Second Law is hopeless. You can't beat it. That's how powerful this idea is.

Macroscopic

Kelvin Statement: perfect engines can't exist

William Thomson, Lord Kelvin (1824–1907) Ireland–Scotland

It is impossible, by means of inanimate material agency, to derive mechanical effect from any portion of matter by cooling it below the temperature of the coldest of the surrounding objects since any restoration of this mechanical energy without more than an equivalent dissipation is impossible.

A Universal Tendency in Nature to the Dissipation of Mechanical Energy

It is impossible to produce work in the surroundings using a cyclic process connected to a single heat reservoir.

It is impossible, by means of inanimate material agency, to derive mechanical effect from any portion of matter by cooling it below the temperature of the coldest of the surrounding objects.

A transformation whose only final result is to convert heat, extracted from a source at constant temperature, into work, is impossible.

On the Dynamical Theory of Heat, with numerical results deduced from Mr Joule's equivalent of a Thermal Unit, and M. Regnault's Observations on Steam.

Clausius Statement: perfect refrigerators can't exist

Rudolf Clausius (1822–1888) Germany

Heat cannot of itself pass from a colder to a hotter body.

It is impossible to carry out a cyclic process using an engine connected to two heat reservoirs that will have as its only effect the transfer of a quantity of heat from the low-temperature reservoir to the high-temperature reservoir.

"Es existiert keine zyklisch arbeitende Maschine, deren einzige Wirkung Wärmetransport von einem kühleren zu einem wärmeren Reservoir ist."

It does not exist a cyclically operating machine, whose only effect is heat transport from a cooler to a warmer reservoir.

It is impossible for a self-acting machine, unaided by any external agency, to convey heat from one body to another at a higher temperature.

Carnot Statement: no engine can be more efficient than a reversible engine

Sadi Carnot (1796–1832), *Réflexions sur la puissance motrice du feu sur les machines propres a developper cette puissance.*

Everyone knows that heat can produce motion. That it possesses vast motive-power no one can doubt, in these days when the steam engine is everywhere known.

The question has often been raised whether the motive power of heat is unbounded, whether the possible improvements in steam engines have an assignable limit — a limit which the nature of things will not allow to be passed by any means whatever; or whether, on the contrary, these improvements may be carried on indefinitely.

According to established principles at the present time, we can compare with sufficient accuracy the motive power of heat to that of a waterfall. Each has a maximum that we cannot exceed, whatever may be, on the one hand, the machine which is acted upon by the water, and whatever, on the other hand, the substance acted upon by the heat. The motive power of a waterfall depends on its height and on the quantity of the liquid; the motive power of heat depends also on the quantity of caloric used, and on what may be termed, on what in fact we will call, the height of its fall, that is to say, the difference of temperature of the bodies between which the exchange of caloric is made. In the waterfall the motive power is exactly proportional to the difference of level between the higher and lower reservoirs. In the fall of caloric the motive power undoubtedly increases with the difference of temperature between the warm and the cold bodies....

The motive power of heat is independent of the agents employed to realize it; its quantity is fired solely by the temperatures of the bodies between which is effected, finally, the transfer of the caloric [heat].

The difference between specific heat under constant pressure and specific heat under constant volume is the same for all gases.

The fall of caloric produces more motive power at inferior than at superior temperatures.

Equivalence of Kelvin, Clausius, and Carnot statements.

"Not all processes which are possible according to the law of the conservation of energy can be realized in nature."

Entropy is :

- The degree to which energy is wasted (dispersed)
- A measure of the unavailability of heat energy for work

Clausius 1865, entropie, *en* to contain, *trope* turning or transformation, "transformation-contents" (trophy: enemy turning around, tropic: sun turns around at tropics of Cancer/Capricorn). Energy is a measure of work *en* to contain, *ergon* deed or work (work, liturgy, organ, orgy). The universe has never been more ordered than it was at its inception. The second law gives us "the arrow of time".

1. The energy of the world is constant.
2. The energy of the universe is constant.
3. kThe entropy of the world strives to a maximum.
4. The entropy of the universe is increasing.

$$\Delta S = \frac{\Delta Q}{T}$$

The Heat Death

Newton's Laws of Motion are reversible in time; that is, they do not distinguish between future and past. Imagine a movie of a process. If one can distinguish whether the movie is run forward or backward, then one has identified a special direction of time and the process is called irreversible. If one cannot tell whether the movie is run forward or backward, then one has not identified a special direction of time and the process is said to be reversible. A movie showing the microscopic behavior of a small group of atoms would look the same if it were run forward or backward, but a movie showing the macroscopic behavior of a collection of molecules wouldn't.

Galileo: We can take wood and see it go up in fire and light, but we do not see them recombine to form wood; we see fruits and flowers and a thousand other solid bodies dissolve largely into odors, but we do not observe these fragrant atoms coming together to from fragrant solids.

In the age of the computer, it's quite easy to imagine writing a program that could keep track of a small number of particles in some kind of container (small, at least, when compared to the unimaginably large number of molecules in a mole). A computer program that could keep track of the particles as they careened around inside the container and as they collided into each other; obeying the conservation of momentum and energy at all times. Although it would be exceptionally tedious for a human, computers don't complain when given these kinds of tasks (not yet, anyway). Before the program started, we could assign to our particles any values of speed and direction that might amuse us. After the program had been running for a while we could halt it, save the positions and speeds of all the particles and reverse their directions. After as much time elapsed as we let the program run originally, our dumb particles would then find themselves returned to their original positions (the ones we arbitrarily chose when we started the simulation) but heading in the opposite direction. On the microscopic scale, the laws of classical mechanics are reversible. That is, there is no preferred direction of time. A movie of a few molecules looks the same run forward as it does when run backward.

Now imagine a whole lot of molecules arranged to form a coffee cup filled with hot coffee. Imagine again, that another collection of molecules arranged to form a hand happen to have been given a velocity that sends them on a collision course with the coffee cup. The two molecular assemblies meet. The force of the collision is great enough to propel the cup toward the edge of the table on which it's resting, but not great enough to cause either the hand or the cup to disintegrate. The molecules in the hand, the cup, and the table are given a bit of a jostle; the bonds between in the cup and the table are insufficient to stop its progress; and the cup slides off the edge. After interacting briefly with the air molecules on the way down, the molecules of the cup strike the molecules of the floor and the cup loses its coherence. Large sections of the cup separate from other large sections; collectively obeying the conservation of momentum in the process. The molecules of the water, which aren't held together as strongly, fly off in a more diffuse manner. All these dispersing molecules are eventually stopped by the molecules of the floor. The cup has broken and the coffee has spilled. On the microscopic level, all the molecules of this system have observed the laws of classical mechanics, which are reversible.

Now imagine a whole lot of molecules arranged to form the shards of a cup coated with a film of molecules that together constitute coffee resting on the molecules of the floor. Imagine again that the molecules of the floor, which are constantly in motion, suddenly conspire to move in a manner that thrusts the shards and the liquid together (but leaves the dirt and grime behind) and that these molecules fall together in the right manner such that the liquid is wrapped by the solid shards and that these shards coalesce into a shape capable of containing the moving liquid. The molecules of the floor give one last push together as a unit and propel the cup and coffee up off of the floor at just the right trajectory so that the whole mess arrives at the level of the table where the molecules of a hand meet them moving in the same direction and, like the hand of a baseball catcher's mitt, bring them to a halt on the table. The coffee has unspilled, the cup has unbroken itself, and the whole assembly has come to rest right where it belongs — within my grasp. On the macroscopic level, such an event is impossible. No group of molecules would ever conspire to do this, but there is no physical reason why they couldn't. Molecules in the floor pushing the molecules of a cup aren't fundamentally different in any way from the molecules of a cup pushing the molecules in the floor. A smashed cup of coffee spilled on the floor transforming into a cup of coffee sitting on a table is an impossibility. This is the heart of what became known as the reversibility paradox.

Ludwig Boltzmann (1844–1906) statistical concept of entropy as a logarithmic tally of the number of microscopic states corresponding to a given macroscopic (thermodynamic) state.

Boltzmann sought to derive the Second Law of Thermodynamics from the statistical behavior of a large number of molecules obeying the simple laws of mechanics during collision — namely the conservation of momentum and energy. Such analyses are given the name statistical mechanics. Boltzmann showed that given a collection of molecules with any velocity distribution, after enough time has passed the velocities of the molecules will eventually acquire a continuous normal-like distribution described by James Clerk Maxwell (1831–1879) and that the gas will eventually acquire a uniform temperature. That is, Boltzmann showed that the arrow of time was statistical in origin. A smashed and spilled coffee cup will never be seen to reassemble into a cup of coffee, not because it can't happen, but because it's extremely unlikely that it ever could happen. The Second Law of Thermodynamics is just a way of telling us not to worry about such things. Time runs forward because that is the most likely way for it to run. It's so overwhelmingly impossible for it to do anything else that the universe as we know it will likely not be around long enough for any macroscopic violation of the Second Law of Thermodynamics to occur.

Entropy is

- a measure of disorder
- the number of identical microstates

$$S = k \log w$$

Loose notes.

- Entropy decreases may occur in small systems over short time-scales. In 2002, researchers at Australian National University monitoring a water solution of 100 microscopic latex beads (6.3 µm in diameter) and saw the entropy decrease for periods lasting several tenths of a second.

- The mathematical relation describing the probability of observing microscopic violations of the Second Law is known as the Fluctuation Theorem and was formulated by Denis J. Evans, Cohen, and Morriss at Australian National University in 1993.

- "… every dynamical phase space trajectory and its conjugate time reversed anti-trajectory, are both solutions of the underlying equations of motion."

- On the microscopic scale:
  o All interactions are mediated through *conservative forces*. Energy is never "lost" in the microscopic realm of molecules, atoms, and subatomic particles. A reduction in the kinetic energy of a group of particles will always be exactly balanced by an increase in their potential energy in one of the four fundamental forms. Energy can always be accounted for.
  o All processes are *reversible*. A movie of the atoms in a gas run in reverse would be indistinguishable from one run forward. A movie of a photon scattering off an electron would look no more right or wrong than a movie of an electron scattering off a photon. Such events have temporal symmetry.

- On the macroscopic scale:
  o All interactions between systems include *non-conservative forces*. The mechanical energy of macroscopic systems will always dissipate. Fresh energy must be added to keep them going. The energy is not destroyed, of course, it just gets dissipated as heat.
  o All processes are *irreversible*. A movie of most any common activity would look odd if run in reverse. Broken objects on the ground don't reassemble and then jump back on the table. There is an obvious *arrow of time* to all activities.

- Microscopic systems are characterized by conservative forces and reversible processes.

- Macroscopic systems are characterized by non-conservative forces and irreversible processes.

- Disordered states outnumber ordered states.

- 1876 Josef Loschmidt reversibility argument — Imagine a highly ordered arrangement. Run it forwards, it becomes disordered. Run it backward from this point, it becomes ordered. Both events follow the laws of mechanics. Doesn't the second one disprove the second law of thermodynamics? Answer, no. Let the reversed movie continue running in reverse. The system returns to disorder. Any violations of the second law are only temporary.

1												
1	1											
1	2	1										
1	3	3	1									
1	4	6	4	1								
1	5	10	10	5	1							
1	6	15	20	15	6	1						
1	7	21	35	35	21	7	1					
1	8	28	56	70	56	28	8	1				
1	9	36	84	126	126	84	36	9	1			
1	10	45	120	210	252	210	120	45	10	1		
1	11	55	165	330	462	462	330	165	55	11	1	
1	12	66	220	495	792	924	792	495	220	66	12	1

Entropy of language

In 1950, Claude Shannon estimated the entropy of written English to be between 0.6 and 1.3 bits per character (bpc), based on the ability of human subjects to guess successive characters in text. Shannon estimated the entropy of written English in 1950 by having human subjects guess successive characters in a string of text selected at random from various sources. In one experiment, random passages were selected from *Jefferson the Virginian* by Dumas Malone. The subject was shown the previous 100 characters of text and asked to guess the next character until successful. The text was reduced to 27 characters (A-Z and space). Subjects were allowed to use a dictionary and character frequency tables (up to trigram) as aids. A text written in English can be considered to a certain extent as a random signal composed of a finite number of symbols (mainly the letters of the alphabet). Applying statistical techniques related to information theory, it is possible to compute an estimate of the entropy rate of English, thus enabling optimal compression of English texts or even "simulate" it

THE MYSTERY OF ENTROPY

Forms of Energy

Entropy is a property of systems which are described in terms of work, heat, and temperature differences. The study of such systems is the subject of **Thermodynamics**. In order to attempt to understand the concept of entropy, it is best to begin by clarifying concepts associated with energy.

Energy can be designated by the **symbol E**, and has units of **Joules** (as well as units of calories, electron-Volts, *etc.* — a calorie is equal to 4.186 Joules at 15°C). Energy consists of macroscopic energy and microscopic energy. Notable forms of macroscopic energy are kinetic energy and potential energy. It is possible to measure the kinetic and gravitational potential energy of an object by defining these terms relative to the surface of the Earth. But there is no absolute (non-relative) kinetic or gravitational potential energy of any object having no frame of reference.

The microscopic energy of an object (the energy of an object's molecules and subatomic particles) is called the **internal energy** (designated by the **symbol U**). It is not possible to measure or quantify the total internal energy of an object even in relative terms. Only the change in internal energy (ΔU) is meaningful. Internal energy includes such things as the translational kinetic energy of molecules (corresponding to the temperature of an object), the rotational energy of molecules, the vibrational energy of molecules and the energy of chemical bonds of molecules as well as intermolecular forces and subatomic particle energy. Kinetic energy and potential energy can be quantified relative to a point on the Earth, but there is no reference point against which the many forms of internal energy can be quantified. (A glucose molecule at zero Kelvin will have internal energy.)

A change of energy is the difference between the initial and final states of an object or system, represented by the symbol ΔE.

Mathematically : $\Delta E = E_{final} - E_{initial}$

According to the First Law of Thermodynamics, energy is neither created nor destroyed (*i.e.*,the net energy of the universe never changes), which means that the total ΔE must always be zero. But forms of energy can change, such as Kinetic Energy (**KE**) and gravitational Potential Energy (**PE**). If an apple falls from a tree, the gravitational potential energy of the apple decreases while the kinetic energy of the apple increases as the apple falls. When the apple hits the ground, kinetic energy of the apple becomes zero, and the internal energy of the apple and Earth (totalled for both) increases by an amount equal to the decrease in kinetic energy.

Mathematically: $\Delta E = \Delta PE + \Delta KE + \Delta U = 0$ (**First Law of Thermodynamics**)

As the apple falls there is a decrease in gravitational potenial energy (ΔPE is negative) that exactly equals the increase in kinetic energy (the positive value of ΔKE), which nets to zero. Until the apple hits the ground, ΔU is zero. After the apple hits the ground there is no more kinetic energy (KE = 0) and the decrease in potential energy (negative ΔPE) is exactly equal and opposite to the increase in internal energy (positive ΔU).

For a gas, there is a form of energy represented as the product of pressure (**P**) and volume (**V**), which is represented as **PV**. The PV energy in a balloon will increase as the temperature or number of molecules in the balloon increases: the volume of the balloon will increase, the pressure in the balloon will stay the same, and the product of pressure and volume will increase. If a gas is in a sealed metal container with rigid walls instead of in a balloon, the PV energy in the container will still increase as the temperature or number of molecules in the container increases: the pressure in the container will increase, the volume of the container will stay the same, and the product of pressure and volume will increase. PV energy is independent of internal energy, but is dependent upon Temperature (**symbol T**) and the number of molecules in the system for which PV energy is being quantified. Unlike other forms of energy, PV energy can be quantified in absolute (non-relative) terms.

Mathematically:

$$PV = nRT$$

where **n** is the number of molecules (in moles) and **R** is the universal gas constant. (The equation is only exactly correct for an ideal gas, which represents molecules as infinitesimal particles that exert no attractive or repulsive force.)

The sum of internal energy and PV energy defines a form of energy called the **enthalpy (symbol H)**. Enthalpy is thermal energy, *i.e.*, the energy associated with changes in temperature. Enthalpy includes PV energy, which varies with temperature (nRT) and with the internal energy associated with chemical reactions. Although enthalpy is often equated with heat (the term enthalpy comes from the Greek word *enthalpien*, which means "heat"), in Thermodynamics enthalpy is distinct from **heat (symbol Q)** and is distinct from internal energy (**U**).

Mathematically, enthalpy is defined as: **H = U + PV**

Because enthalpy includes internal energy, total enthalpy of a system or object cannot be quantified or measured directly. But enthalpy change (ΔH) can be measured. For an exothermic (energy-releasing) chemical reaction ΔH is the energy generated by the reaction (including the PV term for ambient pressure resistance). ΔH is negative for an exothermic chemical reaction (the enthalpy of the system or object is reduced) whereas ΔH is positive for an endothermic (energy-absorbing) chemical reaction (the enthalpy of the system or object is increased by the amount of energy absorbed).

Although work (**symbol W**) can be quantified in energy units (Joules), work is not a form of energy. Work is a means of transferring energy. Objects or systems have **properties** such as temperature, pressure, mass, volume, energy, *etc.*, but work is not a property. An object can be said to contain energy, but an object cannot be said to contain work. If work is done on an object to increase the object's velocity, the kinetic energy of the object will increase, but the object is not said to contain more (or any) work. Work is simply a process by which the energy of an object can be changed. Work is the product of Force (**F**) and the length (**L**) of the distance over which that force is applied:

$$W = F \times L.$$

A ball being thrown will accelerate in the hand of the thrower over the distance through which the hand forcefully moves. The work accomplished by the force of the thrower's hand increases the kinetic energy of the ball. Gravitational force exerted downward on a book sitting on a table is opposed by an equal upward force from the table, the book does not move, and no work is done by either force. In thermodynamics, work is often the work of compression or expansion by a piston on a gas contained in a fixed volume chamber ("PV work").

Heat (**symbol Q**), like work, can be quantified in Joules, calories or other units of energy. Like work, in Thermodynamics it is incorrect to describe heat as a form of energy. An object does not contain a defined amount of heat, just as an object does not contain a defined amount of work. Like work, heat is a *means* of energy transfer (generally because of a temperature difference). Energy is transferred, heat is not transferred. Heat is simply a process by which the thermal energy of an object can be changed. The transfer of one **calorie** of thermal energy to one gram of water will raise the temperature of the water from 14.5°C to 15.5°C (by definition of calorie). The idea that heat is an entity that flows like a fluid was part of the discredited idea that heat is "phlogiston". Despite the fact that phlogiston theory has long been discredited, the terminology with heat is often abused and with such phrases as "heat content", "flow of heat" and "heat transfer". It is so awkward to avoid these phrases that no attempt is generally made to do so, despite the literal inaccuracy. "Heat" is being used to represent "thermal energy" in these cases, rather than the process of thermal energy transfer.

If a wooden block is pushed along a table-top, the force pushing the block is opposite to the frictional force in the table-top, which results in some of the work being transformed into heat.

The change in enthalpy due to a process is not the same as the change in heat due to a process, because enthalpy is a form of energy, but heat is a means of energy transfer. A given change in the internal energy portion of enthalpy (thermal energy) may involve a mixture of heat and work energy transfers. Only by studying the process can the relative contribution of work and heat to final internal energy be determined.

Mathematically: $$dU = \delta Q + \delta W$$

In words: the change in internal energy is the sum of the energy change due to heat plus the energy change due to work. Although internal energy is the sum of work and heat, the more usual covention is to write the formula as:

$$dU = \delta Q - \delta W$$

because of the original application of thermodynamics to heat engines, which output work from an input of heat. The internal energy change is the difference between the heat added and the work removed from the system. dU differs from ΔU in that ΔU represents a finite change whereas dU represents an infinitesimal change. The infinitesimal change of internal energy is a sum of infinitesimal changes of energy due to heat and work. The symbols δQ & δW are used rather than the symbols dQ & dW to denote the fact that the exact contributions of either the heat or the work term cannot be determined. Only the sum of energy change due to both heat and work can be determined exactly.

The **First Law of Thermodynamics** states that energy is conserved, *i.e.*, energy is neither created nor destroyed ($E = mc^2$ is beyond the scope of Thermodynamics, but total of mass and energy would be conserved). The expression

$$\Delta E = \Delta PE + \Delta KE + \Delta U + P\Delta V = 0$$

is thus a statement of the First Law of Thermodynamics in expressing the fact that although potential, kinetic, internal, and PV energy may change as a result of a process, total energy does not change.

The equation

$$dU = \delta Q + \delta W$$

is also sometimes used as a statement of the First Law of Thermodynamics, despite the fact that it is restricted to internal energy rather than all forms of energy. Neither work nor heat are forms of energy or are properties of objects or systems. But they are processes that can change the internal energy of an object or system. An somewhat poor analogy would be to have a bank balance representing internal energy, an to have work and heat represent cash and checks. A bank balance can be altered by deposits or withdrawals of checks or cash, but neither checks nor cash are a bank balance.

Attempts to Describe Entropy

Entropy (designated by the **symbol S**) is an elusive concept that is defined in terms of other elusive concepts. Entropy quantifies energy not available to do work, but entropy can either increase or decrease in a reversible process, and entropy is not quantified in units of energy. The units of entropy are Joules per Kelvin (J/K). Joules are a unit of energy, work or heat, but in the context of the units of entropy, Joules quantifies only heat. So the units of entropy can be described as heat divided by temperature. Like mass, energy or volume, entropy is a property of an object or system.

Entropy is always associated with the heat form of energy transfer. In general, the entropy of an object or system can increase or decrease, associated with thermal energy being transferred into or out of an object or system ("flow of heat" in or out). Although the entropy of an object or system can increase or decrease, the entropy of the universe as a whole can never decrease. As stated in the **Second Law of Thermodynamics**, for any process, the entropy of the universe will either stay the same or increase, but never decrease. Only for a reversible process can the entropy of the universe remain unchanged. For every irreversible process (*i.e.*, every real process) the entropy of the universe increases. The direction of heat transfer is the same as the direction of entropy transfer. When heat is transferred out of a system, entropy is transferred out of the system. When heat is transferred into a system, entropy is transferred into the system.

The following are equivalent statements of the **Second Law of Thermodynamics**.

The Second Law according to Lord Kelvin (William Thompson):

There can be no engine, which when operating in cycles, the sole effect of which is pumping energy from one heat-reservoir and completely converting it into work.

Essentially: heat cannot be converted entirely into work.

The Second Law according to Rudolf Clausius (first formulation):

There can be no process the sole result of which is a flow of energy from a colder to a hotter body.

The Second Law according to Rudolf Clausius (second formulation):

In any spontaneous process, occurring in an isolated system, the entropy never decreases.

Mysteriously, the property entropy is calculated by dividing the non-property heat by the property temperature.

Mathematically: $dS = (\delta Q_R)/T$ or $S = \int(\delta Q_R)/T$

Q_R is thermal energy transfer that is Reversible.

The conversion of one mole of ice into water is a reversible process that occurs at constant temperature (about 273 Kelvins) with about 6000 Joules of heat. Thus,

$\Delta S = 6000/283 = 22$ Joules/Kelvins.

Mathematically, the product of Temperature and Entropy is an energy term (a consequence of the fact that entropy is equal to heat, quantified as energy, divided by temperature). For this reason, Temperature multiplied by an infinitesimal amount of Entropy (TdS) can be equated to infinitesimals of internal energy, enthalpy, volume and pressure:

$$TdS = dU + pdV$$

$$TdS = dH - Vdp$$

$$TdS = C_p dT - RT(dP/P)$$

where C_p is the heat capacity at constant pressure and **R** is the gas constant. "Heat capacity" is the amount of heat required to raise temperature a given amount. One calorie of heat will raise the temperature of one gram on water by 1°C at 15°C. C_p is "entropy per degree".

Reversible and Irreversible Processes

The entropy change of any process will be the sum of the entropy transfer (into or out of a system) plus the entropy production of the process. The amount of entropy produced by a process is a measure of the irreversibility of the process. An irreversible process accomplishes the maximal amount of work from a given amount of heat. But in every real process that extracts work from heat, heat is lost, resulting in an increase in the entropy of the universe (associated with heat that can no longer do work).

Mathematically : $dS = (\delta Q_R)/T + \sigma$

In words: the change in entropy is the sum of (reversible) entropy transfer $[(\delta Q_R)/T]$ plus entropy production $[\sigma]$.

Another way to attempt to understand entropy is to look at how the concept is used. The Second Law of Thermodynamics states that for any process, entropy in the universe either increases or does not change, but never decreases (although the entropy of an isolated system can decrease by contact with another system). Only if a process is reversible does the entropy of the universe not change.

What are examples of irreversible processes?

- Heat travels from a cold body to a warm body, but the reverse does not happen
- Pushing a wooden block along a table-top converts work (kinetic energy) into friction (heat), but the reverse does not happen (friction/heat does not cause a block to move along a table-top)
- If two unconnected chambers — one of which contains a gas and the other of which is a vacuum — are connected, the gas will flow into the vacuum chamber, but a gas will never contract from one chamber to leave a vacuum in the other chamber
- A hot object placed in a cool environment will cool to the temperature of the environment, but an object will never spontaneously absorb heat from an environment to become warmer than the environment
- On Earth, an apple that loses its connection to a tree will fall to the ground, but an apple will never spontaneously (in the absence of another force) rise from the ground to the level of a tree branch

Entropy provides a means of quantifying irreversibility. In an irreversible process the amount of entropy of the universe increases. The amount of entropy increase in the universe associated with an irreversible process corresponds to the amount of energy that is no longer available to perform work.

The entropy of an isolated system increases as the state of equilibrium is approached. Equilibrium corresponds to maximum entropy (maximum disorder). Systems spontaneously proceed toward an equilibrium state (maximum entropy).

Reversible and Irreversible Cycles

The Kelvin-Plank version of the Second Law of Thermodynamics states that it is impossible for an engine working in a cycle to covert heat entirely into work. For example, heat (Q_H) could flow into a chamber and cause pressure in the chamber to increase enough to move a piston. But in a real process, the energy corresponding to the amount of work (W) performed in moving the piston will always be less than the energy corresponding to the amount of heat (Q_H) that caused the gas to expand. Some heat (Q_C) will have been lost. The amount of entropy increase in the universe (ΔS_u) corresponds to Q_C divided by the temperature (T_C) of the lowest temperature heat sink to which the lost heat flowed.

The most efficient processes will be those that occur with a minimal increase in entropy. Thus, imaginary reversible processes (cycles) that involve no entropy production provide a quantifiable theoretical standard for optimal performance that can be compared to real processes (real cycles, real engines).

What is an example of a reversible process? There are no reversible processes in the real world, but physicists can fantasize a reversible process in the same way as they can fantasize a block sliding on a frictionless surface forever with an unchanging velocity.

A **Carnot Cycle** is an imaginary reversible process. A Carnot Cycle consists of two adiabatic work processes and two isothermal heat transfer processes.

An **adiabatic work process** is a work process that occurs without any heat transfer. There can be no entropy change in an adiabatic process because entropy change is always associated with a heat transfer. Therefore, adiabatic work processes are reversible and involve no entropy change. In a graph of Pressure versus Volume (PV graph), pressure drops most steeply as an adiabatic expansion begins, and drops increasingly less steeply as the expansion proceeds. Between two points on a PV graph, there can be many different possible adiabatic expansion curves, but all those curves will have the same entropy on each point of the curve.

An **isothermal heat transfer process** is a heat transfer process that occurs without any temperature change. In the Carnot Cycle, one isothermal heat transfer occurs with a hot reservoir and the other isothermal heat transfer occurs with a cold reservoir. The entropy *increase* (ΔS_H) that occurs with the isothermal heat transfer (Q_H) at the hot reservoir (T_H) is exactly equal to the entropy *decrease* (ΔS_C) that occurs with the isothermal heat transfer (Q_C) at the cold reservoir (T_C). Thus, the entropy increase from the hot reservoir and the entropy decrease to the cold reservoir results in no net entropy change for the Carnot Cycle.

Mathematically: $\quad \Delta S_H = + |Q_H|/T_H$ (hot reservoir)

and $\quad\quad\quad\quad\quad\quad \Delta S_C = - |Q_C|/T_C$ (cold reservoir)

Note that negative *change* in entropy means that entropy is decreasing. Entropy is *never* negative.

Carnot Power Cycle

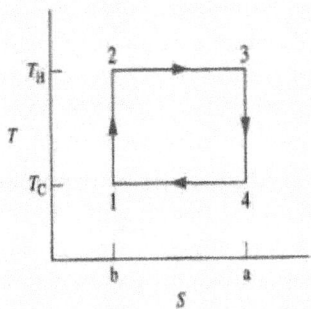

In sum, the adiabatic work processes are reversible and involve no entropy change. The isothermal entropy increase and decrease associated with the two heat transfers are equal and opposite — canceling each other out to give no net entropy change. So the final entropy change after completion of one Carnot Cycle is zero, and the Carnot Cycle is reversible. Four processes in the Cycle occur which cause a change of state between four states (1,2,3,4), which can be represented as points on a PV (Pressure-Volume) graph. The four processes in the Cycle can be graphed as curves which depict the pressure and specific volume changes as each process proceeds between states. The Cycle can also be graphed in a similar manner on a TS (Temperature-Entropy) graph, which results in vertical or horizontal straight lines for each process connecting two states.

A **Carnot Engine** is an imaginary reversible engine that is based on the Carnot Cycle. Gas expands and is compressed in a chamber adjoining a frictionless piston. An equal amount of work is done by and on the piston. An equal amount of heat is transferred in as is transferred out of the chamber.

Carnot Engine

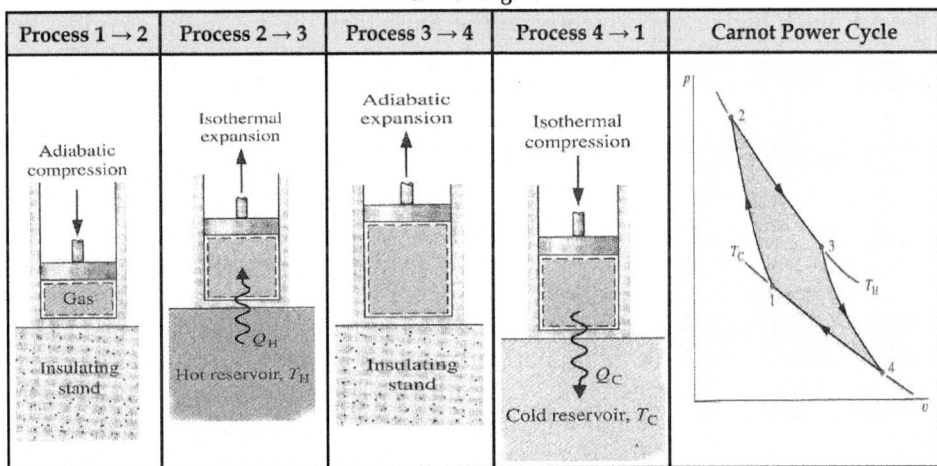

Process 1 → 2	Process 2 → 3	Process 3 → 4	Process 4 → 1	Carnot Power Cycle

- **Process 1 → 2 (reversible adiabatic compression)**: An external force pushes on the piston while the system is thermally insulated (no contact with a heat source or sink). Thus, as the gas is compressed by the piston, the temperature rises from T_C to T_H. [$T = (PV)/(nR)$ and the pressure increases faster than the volume decreases.]

- **Process 2 → 3 (isothermal expansion)**: Heat (Q_H) flows into the chamber very slowly causing the gas to expand and do work on the piston without a change of temperature (T_H). [$T = (PV)/(nR)$ and the pressure decreases at the same rate as the volume increases.]

- **Process 3 → 4 (reversible adiabatic expansion)**: The gas continues to expand and do work on the piston while the system is thermally insulated (no contact with a heat source or sink). Thus, as the gas expands and does work, the temperature drops from T_H to T_C. [$T = (PV)/(nR)$ and the pressure decreases faster than the volume increases.]

- **Process 4 → 1 (isothermal compression)**: An external force pushes on the piston, but instead of temperature rising due to the compression, heat (Q_C) is allowed to flow out of the chamber very slowly causing the gas to contract without a change of temperature (T_C). [$T = (PV)/(nR)$ and the pressure increases at the same rate as the volume decreases.]

The final process returns the Carnot Engine to the original position of the piston and the original temperature, completing the cycle.

Process 2 → 3 and process 4 → 1 might seem especially mysterious insofar as these steps involve a heat flow without a temperature difference. How can heat flow if there is no temperature difference to drive the process? What is the motive force? Actually, heat can flow from one body to another at the same temperature if one of the bodies is undergoing a phase change (gas to liquid or liquid to solid) because heat is absorbed or released during a phase change, even though temperature does not change. Conceivably, the chamber at temperature T_H could be in contact with a vast thermal reservoir undergoing a gas to liquid phase transition (releasing heat into the chamber at T_H). The material in the hot reservoir would have a much higher vaporization temperature than the gas in the chamber that is absorbing heat. The chamber at temperature T_C could be in contact with a vast thermal reservoir undergoing a solid to liquid phase transition (absorbing heat from the chamber at T_C).

In process 4 → 1, for example, an infinitesimal amount of compression would tend to cause an infinitesimal rise in temperature. But because the compression chamber is not insulated, but is imagined to be in thermal contact with a huge (infinite) reservoir at T_C, the infinitesimal temperature rise in the compression chamber causes an infinitesimal flow of heat (energy) out of the compression chamber, thereby keeping the temperature in the compression chamber at T_C. An infinite number of such infinitesimal steps result in a finite isothermal compression in which temperature remains at T_C. A problem with this is that an infinitesimal heat flow at an infinitesimal temperature difference should correspond to an infinitesimal irreversibility. If an infinite number of infinitesimal compressions can sum to a finite compression, then an infinite number of infinitesimal irreversibilities should sum to a finite irreversibility. But through the magic of thought experiment it is claimed that infinitesimal compressions produce equilibrium states rather than infinitesimal non-equilibrium states, that these infinitesimal equilibrium states are reversible, and sum to a finite reversible state.

CALCULATING ENTROPY

Example

The entropy change when one mole of water boils to form one mole of gaseous water at a constant temperature of T = 100°C = 373 Kelvins can be calculated from the enthalpy of vaporization (ΔH_{vap} = 40.67 KiloJoules/mole):

$$\Delta S_{vap} = \Delta H_{vap}/T$$
$$= [\text{mole X } (40.67 \text{ kJ/mole})(1000 \text{ Joules/kJ})]/373 \text{ Kelvins}$$
$$= 109 \text{ J/K}$$

Example

For the reversible isothermal expansion of one mole of an ideal gas from V_1 = 1.0 liter to V_2 = 2.0 liters the increase in entropy will be:

$$\Delta S = Q_{rev}/T = nR \ln (V_2/V_1)$$
$$= \text{mole X } [8.3 \text{ Joules/(mole X Kelvins)}] \text{ X } \ln (2.00)$$
$$= 8.3 \text{ X } 0.67 \text{ Joules/Kelvins}) = 5.7 \text{ J/K}$$

Example

Two containers at different temperatures each containing 1.000 kilograms of water are brought into thermal contact. One container is just below boiling temperature (T_1 = 100°C = 373 Kelvins) and the other container is just above freezing (T_2 = 0°C = 273 Kelvins). Find the net change of entropy when the two containers reach 50°C = 323 Kelvins. The specific heat of water at constant pressure is C_p=4184 J/K

Hot and Cold Water

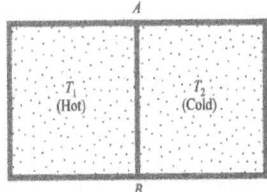

$$\Delta S = \Delta S_1 + \Delta S_2$$
$$= \int_{T1initial}^{T1final} C_P (dT_1)/T_1 + \int_{T2initial}^{T2final} C_P (dT_2)/T_2$$
$$= C_P [\ln (T_{1initial}/T_{1final}) + \ln (T_{2initial}/T_{2final})]$$
$$= 4184 \text{ J/K } [\ln (323K/273K) + \ln (323K/373K)] = 100 \text{ J/K}$$

Third Law of Thermodynamics

The Third Law of Thermodynamics states that the entropy of perfect crystals of all pure elements and compounds is zero at at temperature of zero Kelvin. The Third Law of Thermodynamics provides a standard for the specification of absolute entropy rather than just entropy change. Absolute entropy is written $S°$ The absolute entropy of one mole of some common stubstances at 25°C is given in the following table:

$S°$ in Joules/Kelvins at 25°C		
Substance	$S°$ (Joules/Kelvins)	STATE (solid/liquid/gas)
Diamond (carbon)	2.4	solid
Graphite (carbon)	5.74	solid
Copper (Cu)	33.15	solid
Zinc (Zn)	41.6	solid
Silver (Ag)	42.55	solid
Water (H_2O)	69.94	liquid
Mercury (Hg)	76.02	liquid
Helium (He)	126.1	gas
Argon (Ar)	154.7	gas
Nitrogen (N_2)	191.5	gas
Oxygen (O_2)	205.03	gas
Carbon dioxide (CO_2)	213.6	gas

For solids, $S°$ generally increases with atomic number. Larger, more complex molecules have larger $S°$.

Example

Air is compressed from a pressure of 100 kiloPascals (P_1) to 600 kiloPascals (P_2) causing a temperature increase from 17°C (290 Kelvins) (T_1) to 57°C (330 Kelvins) (T_2). Treating air as an ideal gas and using a heat capacity at constant pressure for air of $C_p = 1.006$ kiloJoules/(kilogram. Kelvins) and the (kilogram) gas constant $R = 0.287$ kiloJoules/(kilogram. Kelvins) gives:

$$\Delta S = C_p \ln (T_2/T_1) - R \ln (P_2/P_1)$$

$$= (1.006 \text{ kJ/kg·K}) \ln (330K/290K) - (0.287 \text{ kJ/kg·K}) \ln (600 \text{ kPa})/(600 \text{ kP}_a)$$

$$= -0.3842 \text{ kJ/kg·K}$$

An alternate calculation, using table values of $S°$ for air of $T_1 = 290$ Kelvins and $T_2 = 330$ Kelvins gives:

$$\Delta S = (S°_2 - S°_1) - R \ln (P_2/P_1)$$

$$= (1.79783 - 1.66802) \text{ kJ/kg·K} - (0.287 \text{ kJ/kg·K}) \ln (600 \text{ kPa})/(600 \text{ kP}_a)$$

$$= -0.3844 \text{ kJ/kg·K}$$

For large temperature differences the second equation will be more accurate.

Gibbs Free Energy

The change in Gibbs Free Energy (ΔG) can be used to predict whether a chemical reaction will occur or not. Gibbs Free Energy is a property of a system that is determined by enthalpy and entropy:

$$\Delta G = \Delta H - T\Delta S$$

Enthalpy change (ΔH) is negative for an exothermic reaction, and if entropy change (ΔS) is positive, both ΔH and $- T\Delta S$ will cause ΔG to be negative. When ΔG for a reaction is negative at constant temperature and pressure, the reaction will occur spontaneously. When $\Delta G = 0$ at constant temperature and pressure, the system is in equilibrium, and no reaction will occur.

The standare Gibbs Free Energy (G°) can be calculated from the standard enthalpy of formation (H_f°) and standard entropy (S°), the values of which can be obtained from tables. As an example, calculate ΔG° for the decomposition of calcium carbonate ($CaCO_3$). At 25°C, for the reaction:

$$CaCO_3 \text{ (solid)} \leftrightarrow CaO \text{ (solid)} + CO_2 \text{ (gas)}$$

$$\Delta H^\circ = H_f^\circ(CaO) + H_{2f}^\circ(CO_2^\circ(CaCO_3)$$

$$= -635.5 - 393.51 - (-1206.9) = 177.9 \text{ KiloJoules}$$

$$\Delta S^\circ = S^\circ(CaO) + S^\circ(CO_{23})$$

$$= 39.7 + 213.6 - 92.9 = 160.4 \text{ Joules/Kelvins}$$

Thus: $\Delta G^\circ = \Delta H^\circ - T\Delta S^\circ$

$$= 177.9 - (298.15)(160.4 \times 10^{-3}) \text{ KiloJoules}$$

$$= 177.9 - 47.82 \text{ KiloJoules}$$

$$= 130.1 \text{ KiloJoules}$$

Because ΔG° is positive, calcium carbonate does not decompose at 25°C, which means that marble and clam shells (which are composed of $CaCO_3$) are stable at 25°C.

ENTROPY AS PROBABILITY

Thermodynamics preceeded the knowledge that matter is made of atoms and molecules. With that knowledge, the absoluteness of the Second Law of Thermodynamics is replaced with an extremely high probability that (for example) a gas will expand irreversibly into an empty chamber.

If two different gas atoms are in a box having two compartments, and there is an opening in the partition separating the two compartments that will allow the atoms to pass through, at any one time there will be only one chance in four that both atoms will be in the left compartment (Probability = ¼).

Two Different Gas Atoms

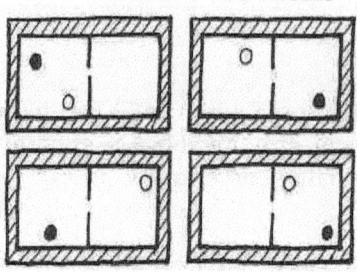

Twelve Gas Atoms

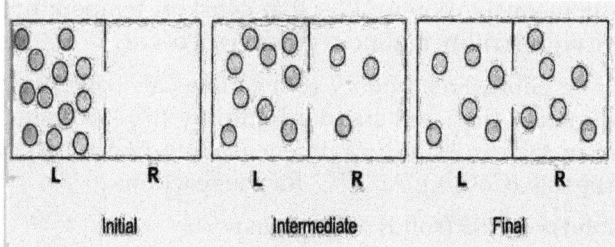

By the same logic, if 12 gas atoms are allowed to equilibrate between two compartments of a box, on average there will be about six atoms in each compartment at the same time — and at any one time there will be only one chance in 24 that all of the gas atoms will be in the left compartment (Probability = 1/24).

Carrying the same logic even further, if billions of atoms are allowed to equilibrate between two compartments, on average about half the atoms will be in each compartment, and the probability is virtually zero that at an one time all of the atoms will be in the left compartment. The distribution of atoms into their most probable arrangement corresponds to an increase in entropy.

Mixing Hot and Cold Gas Atoms

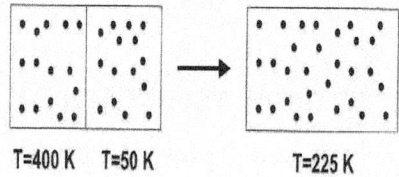

T=400 K   T=50 K   T=225 K

Similarly, if two compartments each contain a gas at different temperatures (T_{hot} = 400 Kelvins and T_{cold} = 50 Kelvins) and the insulating partition separating the compartments is removed, the gas atoms will mix to result in an average temperature ($T_{average}$ = 225 Kelvins) and a net increase in entropy. Maxwell's Demon was an elf imagined by James Clerk Maxwell that could violate the Second Law of Thermodynamics by only allowing fast atoms to pass through a partition when going to the left side, and only allow slow atoms to pass through a partition when going to the right side. Putting energy into a system can decrease the entropy of the system, but only by increasing the entropy of the universe. (Presumably Maxwell's Demon would be burning calories during his labor.)

Mixing Different Atoms

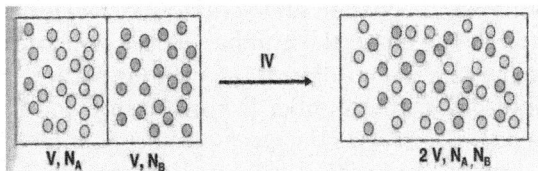

Nonetheless, if a partition separates two different gases (Helium blue and Argon red, for example) at the same temperature and pressure, when the partition is removed the two kinds of atoms will mix, **but the entropy does not increase**. For this reason (among others), the equating of entropy with "disorder" or "mixing" is mistaken.

Coin Hidden in One of Eight Boxes

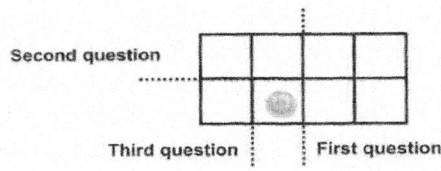

If a coin is hidden in one of eight boxes, there is a 1/8 chance of guessing which box the coin is in on the first guess. After guessing four boxes there will be a 50% chance that the coin has been located. But if the strategy of binary search is used, the coin can be definitely located in only three guesses. The first guess will be "Is the coin in the left half of the group of boxes"? The second guess will be "Is the coin in the lower half of the remaining boxes"? The third and final guess, which locates the coin, will be "Is the coin in the left half of the remaining boxes"? The coin was located in $\log_2 8 = 3$ guesses.

Coin Hidden in 16 or 32 Boxes

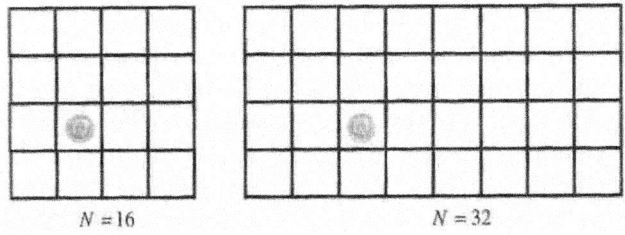

By the same logic, a coin hidden in one of 16 boxes can be located with a binary search in $\log_2 16 = 4$ guesses and a coin hidden in one of 32 boxes can be located in $\log_2 32 = 5$ guesses.

The Austrian physicist Ludwig Boltzmann defined entropy with can the equation:

$$S = k_B \log \Omega$$

where k_B is the Boltzmann constant and Ω quantifies the total number of microscopic states occupied. The larger the number of occupied microscopic states (*i.e.*, the larger the number Ω). The number Ω is a unitless natural number (0,1,2,3,...). In terms of gas molecules, the number Ω corresponds to all the positions and momenta ("momentums") of all of the gas molecules. Geometrically, the possible microscopic states can be represented as quantized space and momentum, where each quantized position is represented as a tiny cube of three spatial dimensions and each quantized momentum is represented as a tiny "momentum cube" in three momentum dimensions. Combining spatial and momentum dimensions gives a six-dimensional "phase space" containing a finite number of six-dimensional cubes. The very least probable arrangement corresponds to all of the gas molecules being in a single six-dimensional cube of phase space, *i.e.*, $\Omega = 1$ and $S = 0$. The most probable arrangment will maximize both Ω and$S$ — which will be the equilibrium condition.

k_B represents the universal gas constant R per molecule of gas. Thus, the units of k_B (J/K) times the natural logarithm of Ω (a unitless natural number) gives the units of S, namely, J/K. By the logic of information theory alluded to above, the base of the logarithm would be 2, *i.e.*, $\log_2$. In practice, Ω cannot be determined, the constant k_B is a "fudge factor" to make the units come out correctly. The equation $S = k_B \log \Omega$ cannot be proven and cannot be used to calculate either absolute entropy or changes in entropy — it is a conceptual equation rather than an equation that can be used to do actual calculations. Part of the problem with 19th century physicists formulating thermodynamics without understanding that matter is made of atoms, is that they did not realize that temperature is kinetic energy of atoms. If temperature had been defined in units of energy, the "fudge factor" k_B would not be needed.

Simple Examples of Entropy

Irreversibility of a process in a limited system need not mean absolute irreversibility of that process in the context of the universe (or wider system). If a hot branding iron is removed from a fire and placed on a rock at normal atmospheric temperature, the branding iron will cool as heat dissipates into the environment from the branding iron. For the limited system of the branding iron, the rock and the air in the vicinity of the branding iron, an irreversible process has occurred. But the cooling of the branding iron is not irreversible in the sense that the branding iron can easily be put back in the fire. If the branding iron is heated to the same temperature it had when it was first removed from the fire, the entropy the branding iron lost during the cooling will be restored to the level of the entropy the iron had when it was first removed from the fire. An irreversible process is still occurring with the fire, but that is outside of the system that is defined to only include the branding iron, the rock and the air in the vicinity of the branding iron. Entropy of the smaller system decreases only as a result of entropy of the larger system increasing.

Similarly, an apple falling from an apple tree represents and irreversible process within the limited system of the apple and the tree. But if a person is added to the system, the person can lift the apple from the ground and place it on a branch of the apple tree. Fat and glucose in the muscle of the person is consumed in an irreversible process that nonetheless provides energy to reverse the process of the apple falling from the tree. Again, entropy of the smaller system decreases only as a result of entropy of the larger system increasing.

A vessel of water placed in a very cold environment will freeze. In forming ice, water molecules organize themselves into a crystal lattice. Although the disorder of the water molecules has decreased, the entropy of the universe as a whole has increased because of the dissipation of heat from the vessel to the cold environment.

FIRST-LAW AND SECOND-LAW CONSERVATION (PROCESS) EFFICIENCIES

Energy Flow in a Diesel Engine

When an engine burns fuel it converts the energy stored in the fuel's chemical bonds into useful mechanical work and into heat. Different types of fuel have different amounts of energy, but in any given gallon or liter of fuel there is a set amount of energy. That's all there is, there ain't no more. No magic devices or cow magnets strapped to the fuel line will change that.

The conservation of energy principle defined by the first law of thermodynamics says that when all of the fuel's energy is released by burning in the engine's cylinders it doesn't disappear. The total quantity of energy stays the same and must be accounted for. In the case of the diesel engine shown below it either becomes thermal energy (heat) or mechanical energy (work). For every 100 units of fuel energy that is burned in the engine a hundred units of converted energy has to end up somewhere. It doesn't disappear.

The picture below shows one example of where the energy goes. The example assumes the engine is operating in what we call "steady state".

ENERGY BALANCE IN A BIG EFFICIENT DIESEL ENGINE

THERMAL OR HEAT
ENERGY OUT EXHAUST
32 UNITS PER HOUR

HEAT RADIATED
TO SURROUNDINGS
1 UNIT PER HOUR

100 UNITS OF ENERGY IN
PER HOUR

HEAT ENERGY OUT
IN COOLING WATER
21 UNITS PER HOUR

USEFUL WORK OUT
41 UNITS PER HOUR

HEAT ENERGY OUT
IN OIL
5 UNITS PER HOUR

©WWW.FTEXPLORING.COM   THERMAL EFFICIENCY = 41%

100 UNITS OF FUEL POTENTIAL ENERGY INTO ENGINE
100 UNITS OF MECHANICAL & THERMAL ENERGY OUT OF ENGINE

THE CHEMICAL ENERGY STORED IN THE FUEL IS CONVERTED TO MECHANICAL ENERGY AND THERMAL ENERGY. THE TOTAL MECHANICAL ENERGY AND THERMAL (HEAT) ENERGY OUT MUST EQUAL THE ENERGY AVAILABLE IN THE FUEL - FIRST LAW OF THERMODYNAMICS.

If I did the numbers right, the sum of all the energy out will equal the sum of the energy into the engine. That's what the First Law of Thermodynamics predicts. If the numbers don't add up, it's due to human error, not a violation of the laws of physics.

Can Energy Be Consumed?

Sometimes people say energy is "consumed". They say something like, "That power plant consumes 80 bugazillion gaggledorks of energy each day". Okay, I never heard that exact example, take it as an analogy.

These people are to be pittied, not condemned. They just haven't read this website yet. Unlike you, they don't understand about the first law of thermodynamics.

You know by now that energy is never consumed. It doesn't go away. It is conserved. It is constant. It only changes or moves. At the end of every process or happening the total amount of energy is unchanged. It did not go away, dissappear, or change amount. It may have changed from chemical to kinetic energy, or electrical energy to thermal, but the total amount stayed the same. *It was not consumed!*

But alas, something does sort of get consumed. That may not be quite the correct word. Something that we engineers call *availabilty* or *exergy* or just plain ol' *usefulness*.

"Call me less useful," said Energy after it was converted.

The exciting Second Law of Thermodynamicsexplores this concept. Energy does get less useful as it moves through its changes.

Notice that for every 100 units or "chunks" of fuel energy burned in the engine, only 41 units of useful energy are delivered to the shaft as rotating useful energy (thermal efficiency = 41%). That doesn't seem so great, does it? But that's actually very good. Most engines do less than 40 - some don't even make it into the 30's. And the old steam locomotives barely got 5 or 6 units of useful work out of 100 units of wood or coal energy (thermal efficiency of 5% or 6%) they burned. Things have improved.

As the drawing shows, more than half of the energy in the fuel leaves the engine as thermal energy, or "waste heat" as we mechanical engineers like to call it. In most engines 30 percent or more goes right out the exhaust stack in the hot exhaust gas. Friction heat from the parts in the engine rubbing together is absorbed by the engine oil which lubricates the rubbing surfaces to keep the parts from overheating and melting. The rest of the thermal energy flows into the cooling water through the engine cylinders mostly. If the engine has an aftercooler some of the thermal energy goes out there.

Engineers that design and build engines pay close attention to where the energy goes. We carefully measure the temperatures and flows of all the fluids into and out of the engine. This allows us to figure out exactly where the energy is going. What we want is to convert as much of the fuel's energy as possible into useful work. To do that we first have to know where the energy is going. Then we can try to figure out how to convert more into work and less into "waste heat".

Thanks to an understanding of the laws of thermodynamics, modern engines are significantly more efficient than they used to be. But as the second law of thermodynamics tells us, there is a limit to how efficient engines can be, and it is not very close to 100%. Some engines are getting close to the practical limits now.

Converting Work to Heat

We take it for granted these days that work can be coverted into heat and heat into work. But this wasn't really understood by humans until fairly recently. The drawing on the right shows how an early experimenter was able to figure out how to compare mechanical energy to thermal energy.

James Prescott Joule was his name. This simple experiment helped to change the world by helping us to understand the way energy really behaves.

CONVERSION OF WORK TO THERMAL ENERGY

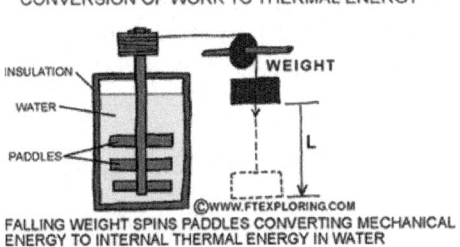

FALLING WEIGHT SPINS PADDLES CONVERTING MECHANICAL
ENERGY TO INTERNAL THERMAL ENERGY IN WATER

The knowledge that mechanical work could be converted to heat and that we could measure and predict how much heat would result from a given amount of work was a huge discovery - a really big deal.

To show our appreciation we named a unit of energy after Mr. Joule. It's called, surprisingly enough, the Joule (J) and in the SI system of units it is equivalent to the mechanical units of work of 1 Newton (force) meter (length). 1 J = 1 N-m.

In the experiment Joule simply attached a weight by pulley and string to some paddles in an insulated container of water. The weight turns the paddles as it falls. The turning paddles do work on the water (they push it all around, "churning it up") equal to the force of gravity on the weight times the distance (L) the weight is pulled down by the gravitational force. It's the formula for work - force times distance. By the time the weight stops, all of it's potential energy at the start of the "fall" has been transferred by the work process into the water (minus a little friction in the pulley and ropes). What happened to the energy? If the first law of thermodynamics is true it had to end up somewhere, it couldn't just dissappear. Joule measured the water temperature and found the temperature had increased. Yup, the water was a little bit warmer because the mechanical work of the paddles had increased the energy level of the water molecules by pushing them around.

DIGRESSIONS & FURTHER EXPLANATIONS SECTION

Energy Stored in Fuel

How much energy is stored in fuel? It depends on the fuel. There's all kinds of fuel. Diesel fuel, gasoline (or petrol as it's called in some countries), methanol in race cars, jet fuel in jets, propane in fork lifts and backyard grills, and some engines burn a gas called natural gas which is mostly made of methane.

It is fairly straight forward with a liquid fuel like diesel to measure the energy content in a sample. A little bit of the fuel is burned in something called a "bomb calorimeter" and the energy released by combustion is measured by the increase in temperature of a surrounding water bath that absorbs the energy.

This is generally called the Heating Value of the fuel. Fuels like diesel and gasoline actually vary quite a bit in composition being generally a hodge-podge of different hydrocarbon molecules. When measuring fuel consumption in an engine lab or doing an energy balance like the one shown in the picture above, it is very important to have an accurate heating value of the fuel being burned, so it is typical to take frequent samples, each of which are sent to the lab to be analyzed. If you don't have an accurate heating value then your value of fuel energy into the engine won't be accurate.

"Energy In" Equals "Energy Out" plus "Energy Stored"

Steady State means the engine is running at a steady constant load and constant speed and is not warming up or cooling down. In this steady state condition the energy flows are also steady and constant.

Sometimes people say the conservation of energy is "Energy In" equals "Energy Out". This is only true for steady state systems. A good example is our engine above. If the engine has just been started after sitting all night and getting cold it will take a while to get warmed up before it reaches steady state. During the warm up period, some of the fuel's energy will go into heating up the cold metal and cold water. The engine's temperature control valves (thermostats) will keep the engine water from flowing to the radiator until it gets warm enough. If we measure the energy flowing out of the engine during the warm up period we would find that less energy is flowing out than is going in. Some energy is being stored in the engine parts and fluids. It is hard to measure this, but if you could you would find that the energy heating up the metal, oil, and water (stored energy), plus the energy flowing out in the oil, cooling water, and exhaust exactly equals the energy in the fuel going in.

Eventually all the fluids and engine parts will reach a constant operating temperature. Then we can say the engine is operating in steady state and the energy out equals the energy in.

Thermal Efficiency

When people talk about the efficiency of an engine, they are usually describing what we engineer's call thermal efficiency. In the case of engines that burn fuel, this is simply the useful mechanical work out of the engine divided by the energy in the fuel burned. Since we know the conversion between mechanical and thermal energy we can relate the thermal energy available in the fuel to the mechanical energy produced by the engine.

Of course, what we are really interested in is power which is how much energy is produced or used in a given amount of time. So when I say my engine is burning 100 units of energy every hour that is actually a measure of power (energy/time). If our engine has a thermal efficiency of 41% then it is producing 41 units of useful mechanical energy every hour. If we are buring 1000 units of fuel energy during the same hour that is 10 times the power. And if the engine burning 10 times the amount of energy every hour also has a thermal efficiency of 41% then it is producing 410 units of energy every hours (410/1000 = 0.41).

With engines like the diesel engine above, we generally define useful energy as the rotating energy delivered at the shaft coming out of the engine. The power produced here is called shaft power or brake power. The effiency then is called brake thermal efficiency. But we can measure the power anywhere we want to, as long as we clearly define it so as not to confuse anyone. For example, if the engine shaft is attached to an alternator we might measure the electricity produced by the alternator. If the alternator has an efficiency of 96% then the power out of the alternator will be 96% of the shaft power from the engine, and we would say the electrical efficiency is about 39%. This is painful, but fair, after all the useful energy to us in this case, is really the electricity coming out of the alternator, not the brake power from the engine going into the alternator.

But the reason we like to define the useful power at the same place on different engines is so that we can compare engines. If one engine has a brake thermal efficiency of 41% and another different engine has a brake thermal efficiency of 38% we know that the engine with the higher efficiency will be more fuel efficient. It will produce the same amount of power while burning *less* fuel. If you are the one paying for the fuel, you will appreciate that.

Thermal Energy *vs.* Heat

If you've read my pages on energy types and heat flow, you know that most thermodynamics text books describe heat as an energy transfer process rather than a type of energy. The same goes for work. By this strict and nit-picky interpretation, work and heat are the two ways that energy can be changed or transferred. Thus they are processes of change rather than types of energy. This strictness is useful to help understand energy better, but I don't really lose sleep over it. It seems that most science books at the high school level and below describe both heat and work as types of energy, so most of the students and teachers visiting my pages are being taught heat and work are types of energy. Some college thermodynamics texts still do discuss heat as a form of energy, while others prefer the term thermal energy or internal energy. I was schooled with the term internal energy which is most descriptive since this type of energy is really made up of all the energies of the atoms and molecules inside a substance. But recently I've decided the term thermal energy is my favorite and so I've started to use it in this website. But, hey, I use all three terms interchangeably, so get used to it. If I just say heat I usually mean thermal energy. If I am describing the process of energy transfer I usually say "heat flow" or "heat flow process".

And if you read and understood all that, you deserve the "hand shake of honor", and maybe even an ice-cream cone.

Entropy and the Second Law of Thermodynamics

The first law of thermodynamics is simply a statement of energy conservation. That is, it states that energy can always be accounted for, that the energy of the universe is a constant - it can be transferred between objects and can change form, but the total doesn't change. But the first law does not preclude things occuring that we know do not occur: A glass of water does not spontaneously separate into ice cubes and warm water even though the energy balance equations used in calorimetry problems would allow it. That is, energy conservation - the first law of thermodynamics - would allow for the possibility that a system in thermal equilibrium could separate into two systems - one at a higher temperature than the other - and that temperature difference could then be used to drive a heat engine to do work. The second law of thermodynamics explains why the universe does not work that way. It articulates the underlying principle that gives the direction of heat flow in any thermal process. The result, of course, fits our everyday experience. The second law states the reason why it is true.

Heat naturally flows from higher temperatures to lower temperatures.

No natural process has as its sole result the transfer of heat from a cooler to a warmer object.

No process can convert heat absorbed from a reservoir at one temperature directly into work without also rejecting heat to a cooler reservoir. That is, no heat engine is 100% efficient.

Carnot Cycle - Maximum Thermodynamic Efficiency in a Cyclic Process

It was observed by Sadi Carnot - a French scientist and engineer trying to improve the efficiency of steam engines in the mid-1800s - that there is always waste heat rejected by a heat engine. And that waste heat limits the efficiency of the engine since energy has to be conserved or accounted for. In trying to understand the limits of efficiency, he stated that in any heat engine in principle there would always be rejected heat (even in an ideal engine) - and the net work done would be the difference between the heat absorbed and that rejected. He then set out to determine the principles that would affect that efficiency. He stated that the most efficient heat engine possible would be one that worked reversibly - an ideal that could never be attained. This would mean that heat transferred into or out of the system (the heat engine) would only occur at constant temperatures - the high or the low temperatures between which the heat engine operated. That is, the system would stay at the temperatures of the reservoirs during those heat transfers - necessary for the process to be reversible since the heat flow could not be reversed to go from the lower to the higher temperature. And furthermore, said Carnot, the maximum conceivable efficiency would be limited by those two temperatures. The most efficient thermodynamic cycle operated between any two temperatures is therefore called a *Carnot cycle*.

The Carnot cycle is a four step process involving two isothermal processes (which are said to be ideal reversible processes) at the temperatures Th and Tc and two adiabatic processes (*i.e.*, without heat transfer) which operate between those two temperatures. In the isothermal steps, there is no change in internal energy and the heat exchanged is equal to the work done. In the two adiabatic processes, there is no heat exchanged. No such system can ever be built - since it is an idealized process (the two isothermal steps being reversible and quasistatic which means, in effect, they occur infinitely slowly). The importance of the process is that it gives an upper limit to the efficiency of any cyclic process between the same two temperatures.

Entropy and the Second Law of Thermodynamics

In trying to synthesize the ideas of Kelvin, Joule, and Carnot - that is, that energy is conserved in thermodynamic processes and that heat always "flows downhill" in temperature - Rudolf Clausius invented the idea of *entropy* in such a way that the change in entropy is the ratio of the heat exchanged in any process and the absolute temperature at which that heat is exchanged. That is, he defined the change in entropy DS of an object which either absorbs or gives off heat Q at some temperature T as simply the ratio Q/T.

With this new concept, he was able to put the idea that heat will always flow from the higher to the lower temperature into a mathematical framework. If a quantity of heat Q flows naturally from a higher temperature object to a lower temperature object - something that we always observe, the entropy gained by the cooler object during the transfer is greater than the entropy lost by the warmer one since Q/Tc.> | Q | /Th. So he could state that the principle that drives all natural thermodynamic processes is that the effect of any heat transfer is a net increase in the combined entropy of the two objects. And that new principle establishes the direction that natural processes proceed. All natural processes occur in such a way that the total entropy of the universe increases. The only heat transfer that could occur and leave the entropy of the universe unchanged is one that occurs between two objects which are at the same temperature - but that is not possible, since no heat would transfer. So a reversible isothermal heat transfer that would leave the entropy of the universe constant is just an idealization - and hence could not occur. All other processes - meaning, all *real* processes - have the effect of increasing the entropy of the universe. That is the second law of thermodynamics.

Entropy is a measure of the disorder of a system. That disorder can be represented in terms of energy that is not available to be used. Natural processes will always proceed in the direction that increases the disorder of a system. When two objects are at different temperatures, the combined systems represent a higher sense of order than when they are in equilibrium with each other. The sense of order is associated with the atoms of system A and the atoms of system B being separated by average energy per atom - those of A being the higher energy atoms if system A is at a higher temperature. When they are put in thermal contact, energy flows from the higher average energy system to the lower average energy system to make the energy of the combined system more uniformly distributed - ie, less ordered. So the disorder of the system has increased - and we say the entropy has increased. But the process of increasing the disorder has removed the possibility that the energy that was transferred from A to B can be used for any other purpose - for example, work cannot be extracted from the energy by operating a heat engine between the two reservoirs of different temperatures. So although energy was conserved in the transfer (the first law), the entropy of the universe has increased in becoming more disordered (the second law) and consequently the availability of energy for doing work has decreased.

The second law of thermodynamics can be summarized in many different statements - and has been by many thermodynamicists in the last century and a half. All of the statements are an attempt to put a reason to the things all of us have observed - that when two objects are in thermal contact, heat always goes from the warmer to the cooler and never the other way. This universal result has probably as many explanations as there are physicists trying to explain it - and is still the subject of serious consideration by some of the best theorists. The difficulty does not lie in what the second law says - or how it should be interpreted - but rather in what the fundamental, underlying reason is for why nature behaves in that way.

Any process either increases the entropy of the universe - or leaves it unchanged. Entropy is constant only in reversible processes which occur in equilibrium. All natural processes are irreversible.

All natural processes tend toward increasing disorder. And although energy is conserved, its availability is decreased.

Nature proceeds from the simple to the complex, from the orderly to the disorderly, from low entropy to high entropy.

The entropy of a system is proportional to the logarithm of the probability of that particular configuration of the system occuring. The more highly ordered the configuration of a system, the less likely it is to occur naturally - hence the lower its entropy.

In the language of entropy, the Carnot cycle still represents the theoretical maximum efficiency in any cyclic process. That is, maximum efficiency would occur if the entropy of the universe did not increase, hence there would be no loss of availability of doing work. But entropy can only remain constant in a reversible isothermal process. So, again, any heat transfer would have to occur isothermally. Therefore the most efficient cyclic process possible involves only reversible isothermal steps and steps in which no heat is transferred - ie, adiabatic. And even in this idealized reversible process in which the entropy of the universe was left unchanged, the efficiency of conversion of heat to work is limited by the two temperatures involved in the isothermal steps.

Based on the ideas of Lord Kelvin, Joule, Boltzmann, Carnot, and Clausius, the first and second laws of thermodynamics can now be restated in two profound sentences:

- The total energy of the universe is a constant.

- The total entropy of the universe always increases.

And these two fundamental principles of nature describe how the universe works.

Heat Engines and the Second Law

The First Law of Thermodynamics is just a restatement of the Work-Energy Theorem or a restatement of the ideas of Energy Conservation. A "simple pendulum" or an "idealized pendulum" would continue to swing **forever!** Think of a movie or a video of a pendulum as it swings back and forth. And if we play the movie or video **backward** we could not distinguish it. And yet thereis an arrow to time! Even pendula slow down. Drop a ball and watch it bounce. Each bounce brings the ball not quite as high as the previous bounce. A movie or a video of such a ball bouncing in the real world **is** different if viewed backward. Ice cubes dropped into a cup of hot tea melt while the hot tea becomes cold.

It would be consistent with the First Law of Thermodynamics if we watched a cup of room temperature tea and found ice cubes forming while the remaining tea became hot and steam arose from it. This is consistent with energy conservation. This scene would require **no violation** of energy conservation. **But we never observe this!** The Second Law of Thermodynamics will explain why (or, at least, begin to!). The Second Law of Thermodynamics deals with the "arrow of time".

Heat Engines

Consider a heat engine that takes some "working substance" -- typically a gas -- through a **cyclic process**. Thermal energy Q_h is absorbed from a source at a **hot** temperature T_h. Net work W is done by the heat engine on its surroundings. Thermal energy Q_c is given to a source at a **cold** temperature T_c.

This can be represented by diagram like this --

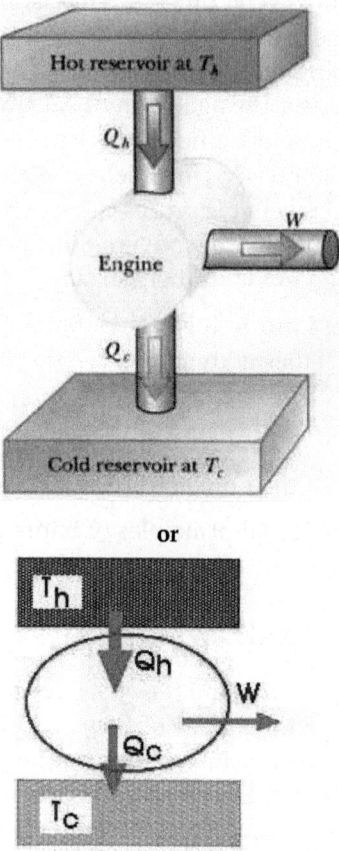

or

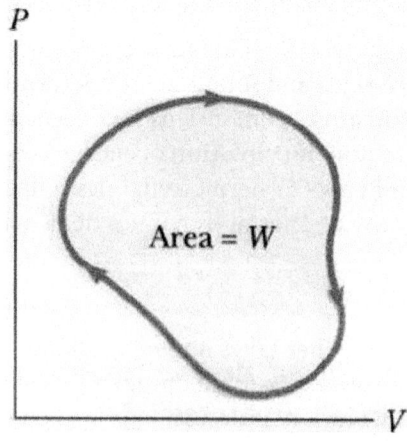

We can also represent this process on a PV diagram. The net work is the area enclosed by the closed curve that represents the cyclic process.

From the First Law of Thermodynamics, we also know that the net Work W is equal to the difference between the heat absorbed Q_h and the heat given up Q_c. That is

$$W = Q_h - Q_c$$

Here, we have taken both Q's to be positive.

The **thermal efficiency** is the ratio of the **work output** to the **input heat Qh**. That is

$$e = \frac{W}{Q_h} = \frac{Q_h - Q_c}{Q_h} = 1 - \frac{Q_c}{Q_h}$$

A typical gasoline engine might have an efficiency of about 20%. A very good diesel engine might even reach 35% to 40%.

Second Law

The **Kelvin-Planck** form of the Second Law of Thermodynamics may be stated as

It is impossible to construct a cyclic heat engine that produces no other effect than the absorption of heat from a reservoir and the performance of an equal amount of work.

We could sketch such an impossible heat engine as

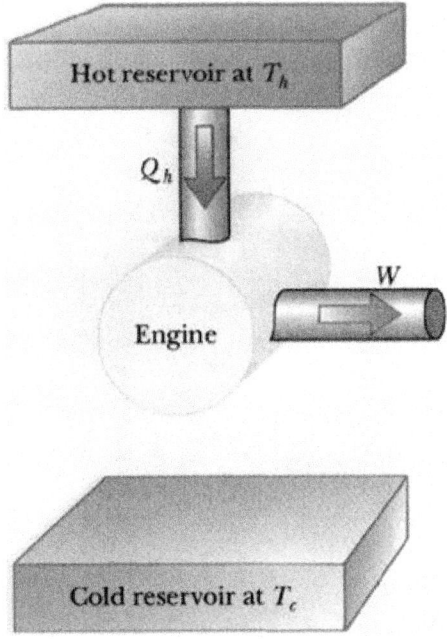

or

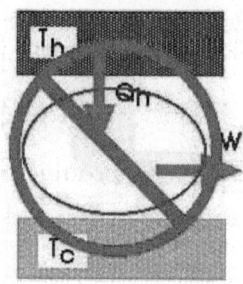

Impossible Heat Engine

This heat engine would then have a thermal efficiency of 1.00. So we could also state the Second Law as

It is impossible to construct a heat engine with an efficiency of 1.00.

Heat Pumps and Refrigerators

A **heat pump** or a **refrigerator** is a device that absorbs heat Q_c from a **cold** temperature and releases heat Q_h at a **hot** temperature because **work W** is done to it from the outside. We could make a diagram of such a refrigerator or heat pump as

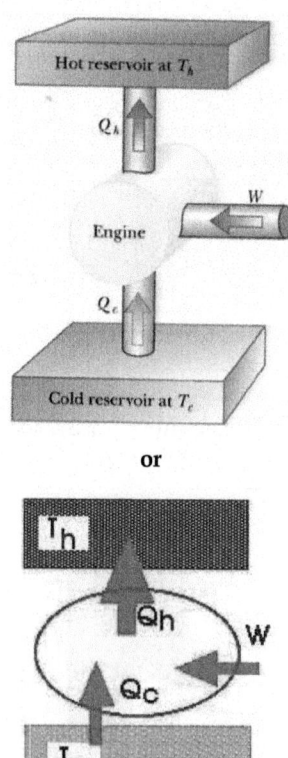

or

Just as with the (ordinary) heat engine, the First Law of Thermodynamics tells us that

$W = Q_h - Q_c$

With a refrigerator or heat pump in mind, there is another form of the Second Law of Thermodynamics, called the **Clausius statement** which says

It is impossible to build a cyclic machine that produces no other effect than to transfer heat continuously from a cold temperature to a hot temperature.

We can make a diagram of this impossible refrigerator,

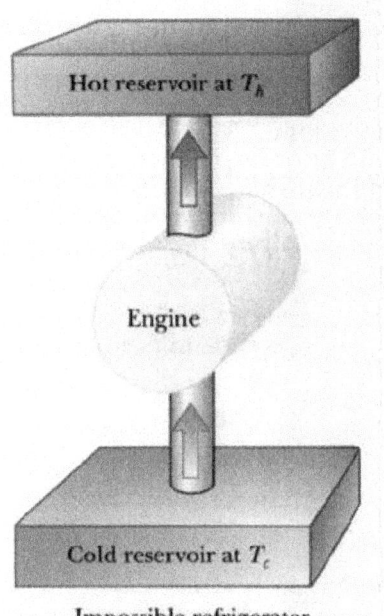

Impossible refrigerator

or

While these two statements of the Second Law sound like they are not connected at all, they are actually **equivalent.** If you could violate one form of the Second Law, you could violate the other form, too. For instance, suppose we could construct a heat engine that had 100% efficiency. It would absorb heat Q_h and turn all of the energy into work W and give out **no heat** Q_c at the cold temperature. We could then **use** this output work as input work for a refrigerator or heat pump.

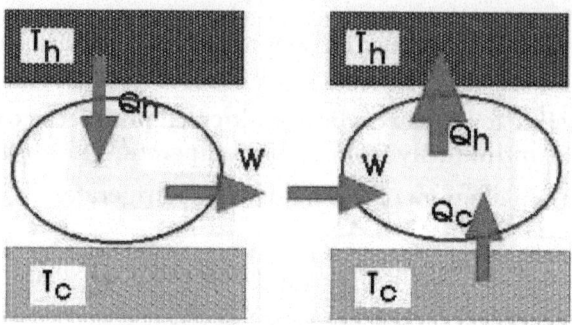

The net result of coupling these two heat engines -- or this impossible, 100% efficient heat engine and an ordinary heat pump -- would be that heat Q_c would be absorbed at the cold temperature and heat Q_h would be expelled at the hot temperature.

And that result is impossible according to the "Clausius form" of the Second Law! If one form of the Second Law is violated, the other will be as well.

Reversible and Irreversible Processes

Drop a glass and watch it break. If we run a video of that **backward** we will surely know the difference. Some processes are clearly **irreversible** -- like the shattering glass.

An example of an **irreversible** process in a "thermodynamic situation" could be a gas and a vacuum separated by a barrier -- like a membrane of aluminum foil -- as sketched here. If the barrier is broken the gas expands to fill the entire container. This sudden expansion into the vacuum does **no work**. The insulation prevents heat exchange. The gas molecules have the same velocity as before the break in the barrier so the temperature remains the same.

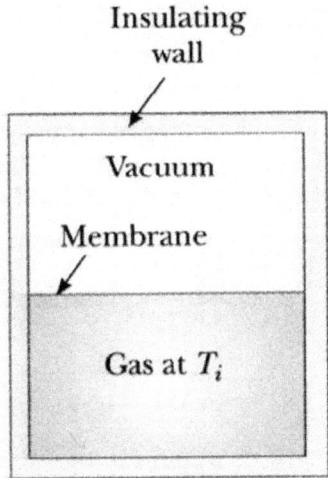

Insulating wall

Vacuum

Membrane

Gas at T_i

While we could argue that **any** or **all** processes are **irreversible**, we can approach **reversible** processes. One example of a **reversible** process could be a **quasi-static** expansion of a gas. We will let it expand slowly enough that all of the gas is always very close to an equilibrium temperature.

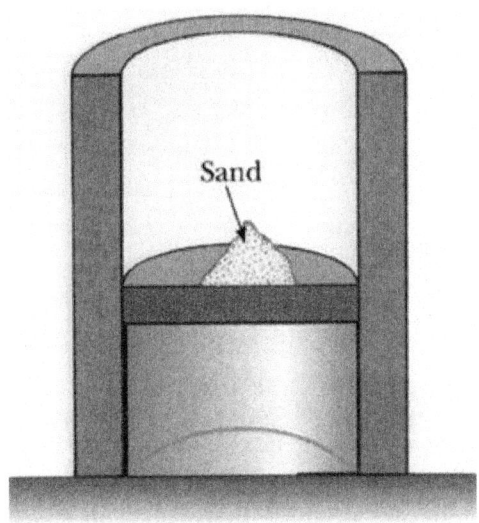

Sand

Heat reservoir

Carnot Engine

The Carnot engine -- or the Carnot cycle -- is important because it describes a heat engine that uses **reversible** processes that can be handled theoretically.

The **efficiency** of a Carnot engine -- or **any reversible** heat engine -- is the greatest that is possible to achieve. Call the efficiency of the Carnot engine e_c.

Then suppose the efficiency of some real heat engine, e_r, is **greater** than that of the Carnot engine,

$$e_r > e_c$$

Then we could use the real heat engine to power a Carnot cycle heat pump. If $e_r > e_c$, then the net result would be the transfer of heat from a **cold** temperature to a **high** temperature. But this violates the Second Law.

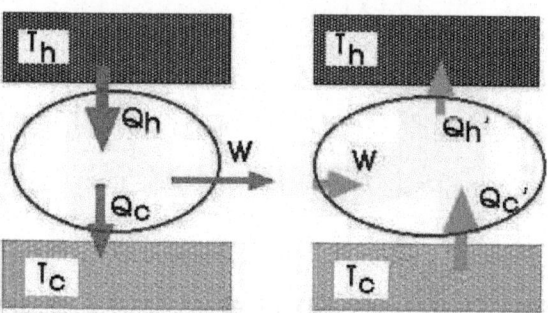

Q_h is the heat absorbed from the **high** temperature by the **real** heat engine and Q_h' is the heat expelled to the **high** temperature by the Carnot cycle heat pump. For this analysis, both Q_h and Q_h' are intrinsicly positive.

$$e_r > e_c$$
$$W/Q_h > W/Q_h'$$
$$1/Q_h > 1/Q_h'$$
$$Q_h < Q_h'$$
$$Q_h' > Q_h$$
$$Q_{h,net} = Q_h' - Q_h$$
$$Q_{h,net} > 0$$

Since we believe the Second Law, that means our assumption that $e_r > e_c$ is **wrong**.

A Carnot engine -- or a Carnot cycle -- is a combination of isothermal expansions and compressions and adiabatic expansions and compressions.

A Carnot engine -- or a Carnot cycle -- is a combination of isothermal expansions and compressions and adiabatic expansions and compressions.

From an initial stat **A**, the gas is placed in contact with the **hot** temperature reservoir (T_h) and expands isothermally (keeping $T = T_h$ = constant) to some state **B**. During this isothermal expansion heat Q_h flows into the gas from the hot temperature T_h.

From state **B**, the gas undergoes an adiabatic expansion to state **C**. No heat is exchanged during this expansion. Expanding an insulated gas means work is done at the "expense" of the internal energy. That means the gas will have a lower temperature. This is the **cold** temperature Tc.

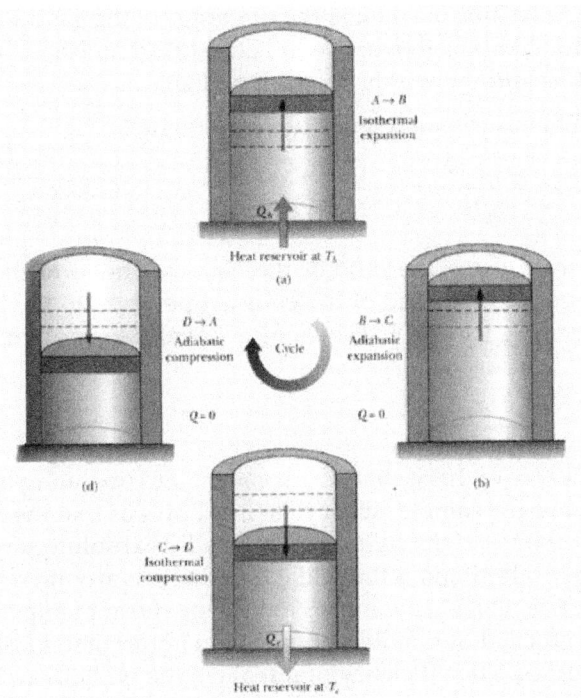

At state **C**, we place the gas in contact with the **cold** temperature heat reservoir (like a large tank of water) and do an isothermal compression to state **D**. In compressing the gas, work is done on the gas by the outside. But the temperature remains constant -- meaning the internal energy U of the gas remains constant. For this to happen, heat Q_c is given out to the cold temperature heat reservoir.

From state **D** we do an adiabatic compression back to state **A**. Remember, "adiabatic" means insulated so there is no heat exchange.

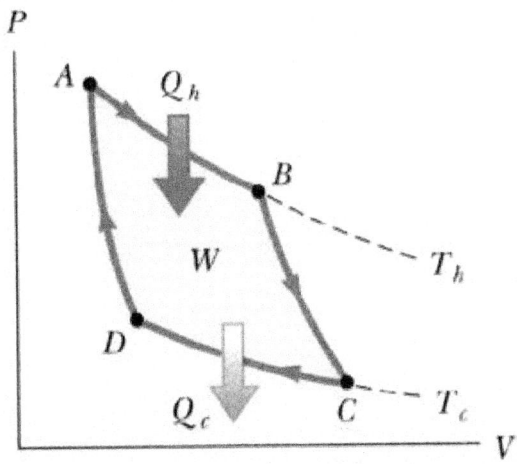

These processes may be seen in the drawings of the gas and piston **or** on the PV diagram immediately above. It will be useful to go back and forth between the two diagrams **and** the words describing them.

We already know the **efficiency** of a heat engine,

$$e = \frac{W}{Q_h} = \frac{Q_h - Q_c}{Q_h} = 1 - \frac{Q_c}{Q_h}$$

It can be shown that the ratio of the heat expended to the heat absorbed, Q_c/Q_h, is also equal to the ratio of the **cold** temperature to the **hot** temperature, T_c/T_h. Then the efficiency of a Carnot engine can also be written as :

$$e = 1 - \frac{T_c}{T_h}$$

The significance of this is that all Carnot engines operating between the same temperatures have the same efficiency. It also shows us that the only way to have an efficiency of 100% is if the cold temperature T_c is absolute zero -- that is, $T_c = 0$ K. For most practical situations the cold temperature is around room temperature. So increasing efficiency usually means increasing the hot temperature Th. That is why most fuel-efficient automobile engines run hotter than most poorly efficient automobile engines. The efficiency of a real engine will always be less than the efficiency of a Carnot engine running betwen the same temperatures.

Chapter 4

LIFE CYCLE ASSESSMENT (LCA)

A systematic set of procedures for compiling and examining the inputs and outputs of materials and energy and the associated environmental impacts directly attributable to the functioning of a product or service system throughout its life cycle.

Life-cycle assessment has emerged as a valuable decision-support tool for both policy makers and industry in assessing the cradle-to-grave impacts of a product or process. Three forces are driving this evolution. First, government regulations are moving in the direction of "life-cycle accountability;" the notion that a manufacturer is responsible not only for direct production impacts, but also for impacts associated with product inputs, use, transport, and disposal. Second, business is participating in voluntary initiatives which contain LCA and product stewardship components. These include, for example, ISO 14000 and the Chemical Manufacturer Association's Responsible Care Program, both of which seek to foster continuous improvement through better environmental management systems. Third, environmental "preferability" has emerged as a criterion in both consumer markets and government procurement guidelines. Together these developments have placed LCA in a central role as a tool for identifying cradle-to-grave impacts both of products and the materials from which they are made.

The "life-cycle" or "cradle-to-grave" impacts include the extraction of raw materials; the processing, manufacturing, and fabrication of the product; the transportation or distribution of the product to the consumer; the use of the product by the consumer; and the disposal or recovery of the product after its useful life.

There are four linked components of LCA:

- **Goal definition and scoping**: identifying the LCA's purpose and the expected products of the study, and determining the boundaries (what is and is not included in the study) and assumptions based upon the goal definition;

- **Life-cycle inventory**: quantifying the energy and raw material inputs and environmental releases associated with each stage of production;

- **Impact analysis:** assessing the impacts on human health and the environment associated with energy and raw material inputs and environmental releases quantified by the inventory;

- **Improvement analysis:** evaluating opportunities to reduce energy, material inputs, or environmental impacts at each stage of the product life-cycle.

In recent years there have been increasing discussions and investigations on environmental impacts resulting from growing consumption of energy and raw materials on one side and emission of solid, liquid and atmospheric waste on the other side. Since the early 1980s, issues such as global warming, ozone degradation in the stratosphere, depletion of natural resources, acidification of water and soil, human and eco-toxicity, *etc.*, have engaged scientists and environmentalists all over the world to develop new methods towards environmental impact assessment. One of the methods being developed for this purpose is the LCA of products.

During the last decade, environmental aspects have become more important and the impacts on nature, consumption of non-renewable materials and energy, and sustainability are important issues concerning design and manufacture of products and their end of life, *i.e.* recycling, burning or landfill.

The new approach to environmental conservation is a more comprehensive one. Sustainable management of natural resources includes both the minimal consumption of materials (renewable or non-renewable) and the protection/conservation of the environment. Technical measures serving to achieve the goal of sustainability are:energy saving, improved use of materials, reuse and recycling, emission control, *etc.* Some of these measures can also have a positive economic effect such as the case of waste paper recycling.

Wood, as a renewable raw material, has been used worldwide for a broad range of end products as well as for renewable energy generation. There are sectors where wood is facing substitution pressure from other materials such as synthetics, concrete, ceramics and glass. This pressure could be reduced if the sustainability of forests and roundwood production were guaranteed, and if consumer awareness of the ecological benefits of wood-based products were enhanced.

The selection of a material for specific end uses strongly depends on its physical and technological suitability, but the cost aspect is also of extreme importance. Due to increased awareness of environmental aspects, certain issues such as low energy demand for producing wood products, possibility of recycling and utilization of waste wood for energy generation at the end of life cycle should be taken into account so to demonstrate the advantage of wood when comparing it with other materials.

LCA is a useful tool for comparing the environmental aspects of specific products as it enables the ecological comparison of two or more products made of different raw materials but used for the same purposes. The results of a comparative LCA study provides useful data for decision-making when selecting environmentally sound fuels, raw materials, products and production processes.

LCA is an approach to study the environmental aspects and potential impacts throughout a product's life from raw material acquisition or production (*e.g.* wood production in the forest) to manufacturing, use, recycling and disposal.

The ISO/EN 14040 defines LCA as a technique for assessing the environmental aspects and potential impacts associated with a product by:

- Compiling an inventory of relevant inputs and outputs of a system;
- Evaluating potential environmental impacts associated with those inputs and outputs; and
- Interpreting the results of the inventory analysis and impact assessment in relation to the objectives of the study.

As a useful tool of decision-making, results of LCA are applied by different groups such as producers' associations, environmental organizations, policy-makers and consumers. The critical consumers are particularly interested in information on the ecological relevance of the commodities and products of short-term and long-term use. For the producer, marketing aspects and ecologically-based optimization of production and processing can be of the highest priority so the critical consumers are satisfied and energy and material costs are reduced.

LCA also gives the possibility to environmental organizations to increase public awareness on environmental aspects and consequently urge policy-makers to provide frameworks for adequate laws and regulations.

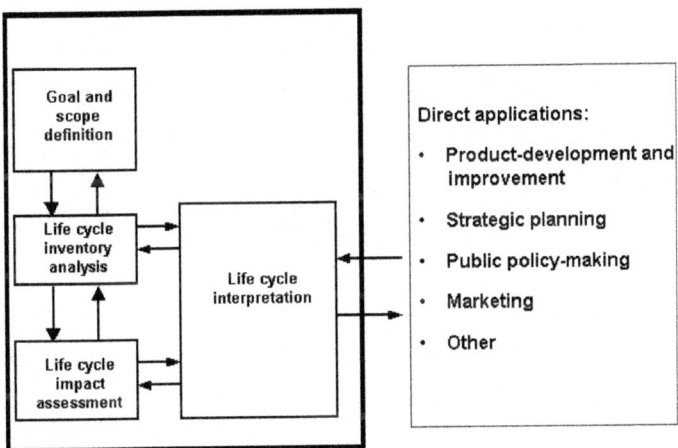

Fig. : Four steps of a life cycle assessment.

Goal and Scope Definition

The goal shall state the intended application, the reasons for carrying out the study and the intended audience. The scope describes among others the function, functional unit, product system to be studied, product system boundaries, allocation procedures, types of impacts and methodology of impact assessment, data and data quality requirements, assumptions and limitations.

The primary purpose of a functional unit is to provide a reference to which the inputs and outputs are related. This reference is necessary to ensure comparability of LCA results. Therefore, the functional unit shall be clearly defined and measurable. For wood products, 1 m³, 1 tonne or 1 m² are usually taken as functional units. Moreover, the moisture content of the product to be studied shall also be specified. For a precise functional unit, oven-dry condition is recommended.

The system boundaries determine unit-processes or life cycle phases to be considered within the framework of an LCA study. The criteria used in establishing the system shall be identified and justified in the scope phase of the study.

The data quality requirement should address time-related coverage, geographical and technology coverage, precision, completeness and representativeness of the data, consistency and reproducibility of the methods used, sources of the data and uncertainty of the information.

Life Cycle Inventory Analysis

The life cycle inventory (LCI) indicates the relevant inputs and outputs of a product system. It includes on the input side energy (electric and thermal) and materials (raw materials, semi-fabricated products, supporting and operating materials) and on the output side products, by-products and releases to air, water and land. Generally, processing machines, buildings, manpower and the use of land necessary for transportation or biological production are not included in LCI otherwise the inventory analysis would become very complex.

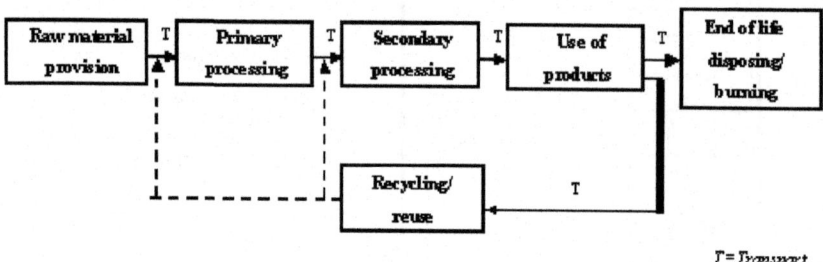

T = Transport

Fig. : Life cycle phases to be considered within the framework of LCA studies.

A complete LCI comprises all phases of a product life cycle. In the case that the system boundaries defined include only one or few life cycle phases (*e.g.* production and use of certain products), the collected data must cover the phases concerned.

When dealing with systems involving multiple products, allocation procedures are needed (*e.g.* multiple products from petroleum refining). In the timber industry there are often products and by-products where products are, for example, sawnwood or furniture and typical by-products are different types of residues which serve as raw material for other products such as particle board, fibreboard or pulp and paper.

Electric and thermal energy, together with renewable and fossil energy, and the relevant energy mix and efficiency, should be taken into account when calculating energy consumption and the resulting emissions.

The life cycle of a timber-based roof. It comprises the phases "roundwood production in the forest", "sawnwood production in the sawmill", "production of roof elements in the carpentry and roof assembly at the construction site", lifetime (using time) including demolition and waste wood utilization (recycling and burning). It is also necessary to consider the transportation between the life cycle phases. The transportation can be analysed either separately for each phase or for the entire life cycle.

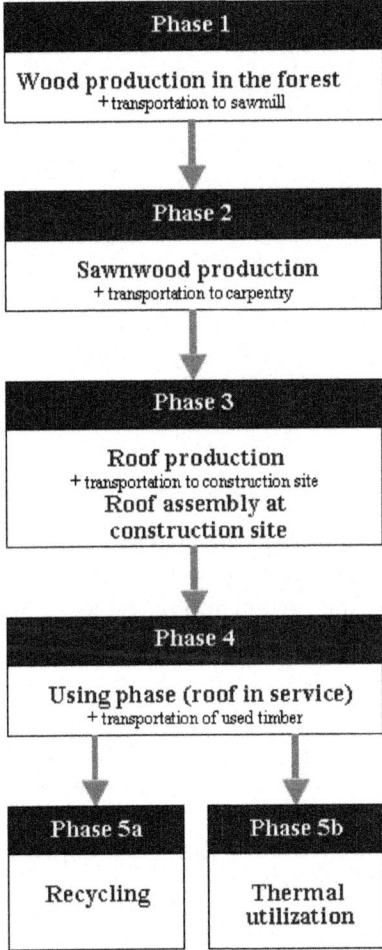

Fig. : Life cycle of a timber-based roof.

A complete LCI includes the energy and material flow for all phases. In this case, the system boundaries are the beginning of roundwood production and the end of life cycle (recycling or burning of waste wood). LCA studies are often confined to a few life cycle phases, but it is possible to reconstruct the entire life cycle by using two ormore studies as literature sources.

For the roof as a timber-based product, the flow of material and energy. The output of one phase (*e.g.* wood production in forest) is the input for the next phase (*e.g.* sawnwood production). However, there are some output materials (*e.g.* thinnings, emissions resulting from fossil fuel) that leave the boundary system. Thinnings as forest output leave the forest along with roundwood. While roundwood remains within the boundary system, thinnings leave the system and are utilized as fuel or as raw material for paper and oriented strand board (OSB) production. Emissions such as CO_2 or NO_x are released to the atmosphere and, therefore, they also leave the boundary system.

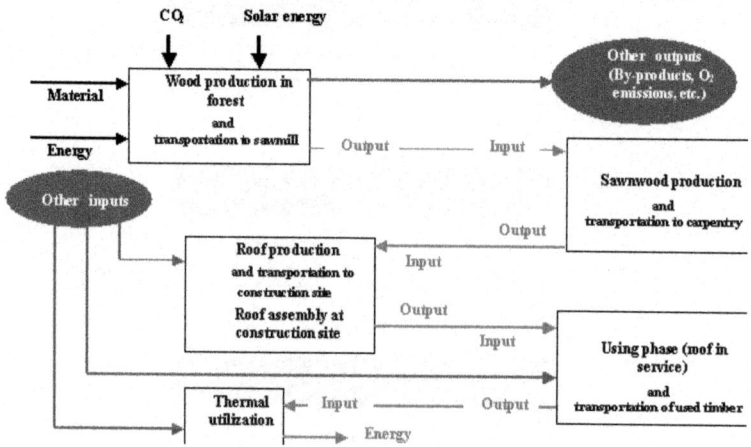

Fig. : Life cycle assessment of a roof construction.

Beside roundwood and energy there are also other inputs and outputs such as operating materials, supporting materials, residues, different types of emissions, *etc.*, that partly remain in the system and partly leave the boundaries. Showing all these materials in one figure leads to confusion, therefore, complete and shows only the main stream.

Life Cycle Impact Assessment

The life cycle impact assessment is aimed at evaluating the significance of potential environmental impacts using the results of LCI. The level of detail, choice of impacts evaluated and methodologies used depend on the goal and scope of the study concerned. For a proper impact assessment the following three steps are generally necessary:

- Classification: assigning of inventory data to impact categories;
- Characterization: modelling of inventory data within impact categories; and
- Valuation: possibly aggregating the results in very specific cases and only when meaningful.

The impact categories can be based on global, regional, local and other criteria and these are described below.

Global impact categories	Regional impact categories	Local impact categories	Other impact categories
• use of resources (renewable and non-renewable), water and land • greenhouse effect/GWP • ozone degradation in the stratosphere • release of persistent toxic substances	• acidification of water and soil • waste disposal/landfill	• human toxicity • eco-toxicity • photochemical ozone creation	• use of land (transportation roads, need of large areas for biological production) • noise • unpleasant smell, *etc.*

Life Cycle Interpretation

According to ISO/EN 14040, interpretation is the phase of LCA in which the findings from the inventory analysis and the impact assessment are combined together. The findings can result in conclusions and recommendations to decision-makers, consistent with the goal and scope of the study. Moreover, the findings should reflect the results of any sensitivity analysis that is performed.

DEFINITION AND MEANING

The International Organization for Standardization (ISO) defines life cycle assessment (LCA) as the following:

"Compilation and evaluation of the inputs, outputs and the potential environmental impacts of a product system throughout its life cycle" (ISO 14040: 1997).

Among the tools that life cycle management (LCM) offers, LCA, or environmental balance, is one of the most comprehensive and high-performance methods. Life cycle assessment is the only method that assesses the environmental impacts of a product or activity (a system of products) over its entire life cycle. It is therefore a holistic approach that takes into account:

- Extraction and treatment of raw materials
- Educational tools
- Product manufacturing
- Transport and distribution
- Product use
- End of life

The main goal of the method is to lessen the environmental impacts of products and services by guiding the decision-making process. For companies, designers, and governments, LCA represents a decision-making aid tool for implementing sustainable development.

Regulated by the ISO 14040 series standards, LCA consists in four distinct phases:

- Goal and scope definition (study model which defines the methodological framework which all other LCA phases must comply with)
- Inventory of all the inputs and outputs related to the product system
- Assessment of the potential impacts associated with these inputs and outputs
- Interpretation of the inventory data and impact assessment results related to the goal and scope of the study.

The following figure illustrates the LCA framework as described by ISO. As shown in the figure, LCA is an iterative process. The choices made during the course of the study can be modified following new information.

Due to time and cost constraints, variants of the LCA method have been formulated according to the guideline principles established by the Society of Environmental Toxicology and Chemistry (SETAC). For example, the boundaries of a system can be limited to certain life cycle stages or to certain impact categories or limited to the main contributors, identified according to expert opinion and experience.

The analysis can also be performed in a strictly qualitative fashion or using « secondary » data only (generic data from the literature or from databases). Such simplifications can affect the accuracy and applicability of LCA results, but can nevertheless allow for the identification of potential impacts and, to a certain extent, their assessment.

Life cycle assessment has become an important tool for the environmental impact assessment of products and materials and businesses are increasingly relying on it for their decision-making. The information obtained from an LCA can also influence environmental policies and regulations.

Life cycle assessment is used for comparing on an environmental basis:

- Various types of similar products for granting eco-labels
- Life cycle stages to enhance the environmental value of a product
- Processes or services for conformity analyses
- Production methods
- Product and process choices

Life cycle assessment can be employed with the aim of finding the most eco-logical way to improve product manufacturing. Consequently, it can be useful as a decision-making tool for new product development, as a guide for the optimization of energy and raw material consumption as well as for the identification of solutions in emission reduction and possible substitution of harmful substances.

Life cycle assessment can also be used for commercial development:

- For obtaining authorizations (in the case of a new product for example)

- For marketing (in the case of a certification or the creation of a marketing image)
- As an aid for informing the public (within the framework of an environmental product declaration (EPD), for example).

A significant number of LCA studies, European for the most part, have already been conducted on several different subjects. The following list is not exhaustive:

- Clothing
- Grocery bags
- Packaging and containers
- Vending machines
- Edible products
- Personal computer
- Diapers
- Paper
- Buildings
- Cars
- Bottles
- Pharmaceutical products
- Sanitary landfill
- Contaminated sites
- Means of transport
- Hotel services
- Doors and windows
- Electric household appliances

Life cycle assessment (LCA) is a tool which offers many advantages:

- LCA is the only tool that examines the environmental impacts of a product or service throughout its life cycle
- LCA is an ISO standardized method
- LCA provides a comprehensive overview of a product or service and avoids simply shifting the source of the pollution from one life cycle stage to another
- LCA can, for example, guide a company's decision-making process (micro-economic level) and help governments define a public policy (macro-economic level)
- LCA challenges preconceived notions by distinguishing between the information that is relevant for objective quantification and the issues that pertain to policies, priorities, and social choices

LCA has however certain limitations:

- The results of an LCA are geographically dependent. Hence, the results of an LCA carried out in Europe cannot be applied to Quebec without taking into account the significant variations related to the geographical context (for example, Québec relies on hydroelectricity while Europe emplys other sources of energy such as nuclear)

- LCA only assesses potential impacts and not real impacts. Hence, it does not provide any information on the consequences of not following regulations or on environmental risks

- The results of two LCAs on a same subject may differ according to the objectives, processes, quality of the data, and the impact assessment methods used. This is why ISO insists on transparency in LCA

- A detailed LCA requires inventory data of all of the elementary processes included within the parameters of the system. Databases, LCAsoftware, and even human resources are required to analyze all the data

Goals and Purpose

The goal of LCA is to compare the full range of environmental effects assignable to products and services by quantifying all inputs and outputs of material flows and assessing how these material flows have an impact on the environment. This information is used to improve processes, support policy and provide a sound basis for informed decisions.

The term *life cycle* refers to the notion that a fair, holistic assessment requires the assessment of raw-material production, manufacture, distribution, use and disposal including all intervening transportation steps necessary or caused by the product's existence.

There are two main types of LCA. Attributional LCAs seek to establish the burdens associated with the production and use of a product, or with a specific service or process, at a point in time (typically the recent past). Consequential LCAs seek to identify the environmental consequences of a decision or a proposed change in a system under study (oriented to the future), which means that market and economic implications of a decision may have to be taken into account. Social LCA is under development as a different approach to life cycle thinking intended to assess social implications or potential impacts. Social LCA should be considered as an approach that is complementary to environmental LCA.

The procedures of life cycle assessment (LCA) are part of the ISO 14000 environmental management standards: in ISO 14040:2006 and 14044:2006. (ISO 14044 replaced earlier versions of ISO 14041 to ISO 14043.) GHG product life cycle assessments can also comply with standards such as PAS 2050 and the GHG Protocol Life Cycle Accounting and Reporting Standard.

Four Main Phases

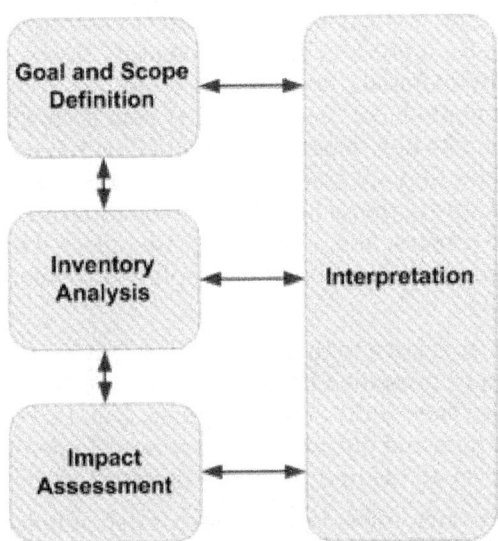

Fig. : Illustration of LCA phases.

According to the ISO 14040 and 14044 standards, a Life Cycle Assessment is carried out in four distinct phases as illustrated in the figure shown to the right. The phases are often interdependent in that the results of one phase will inform how other phases are completed.

Goal and Scope

An LCA starts with an explicit statement of the goal and scope of the study, which sets out the context of the study and explains how and to whom the results are to be communicated. This is a key step and the ISO standards require that the goal and scope of an LCA be clearly defined and consistent with the intended application. The goal and scope document therefore includes technical details that guide subsequent work:

- The functional unit, which defines what precisely is being studied and quantifies the service delivered by the product system, providing a reference to which the inputs and outputs can be related. Further, the functional unit is an important basis that enables alternative goods, or services, to be compared and analyzed.
- The system boundaries;
- Any assumptions and limitations;
- The allocation methods used to partition the environmental load of a process when several products or functions share the same process; and
- The impact categories chosen.

Life Cycle Inventory

Life Cycle Inventory (LCI) analysis involves creating an inventory of flows from and to nature for a product system. Inventory flows include inputs of water, energy, and raw materials, and releases to air, land, and water. To develop the inventory, a flow model of the technical system is constructed using data on inputs and outputs. The flow model is typically illustrated with a flow chart that includes the activities that are going to be assessed in the relevant supply chain and gives a clear picture of the technical system boundaries. The input and output data needed for the construction of the model are collected for all activities within the system boundary, including from the supply chain (referred to as inputs from the techno-sphere).

The data must be related to the functional unit defined in the goal and scope definition. Data can be presented in tables and some interpretations can be made already at this stage. The results of the inventory is an LCI which provides information about all inputs and outputs in the form of elementary flow to and from the environment from all the unit processes involved in the study.

Inventory flows can number in the hundreds depending on the system boundary. For product LCAs at either the generic (*i.e.*, representative industry averages) or brand-specific level, that data is typically collected through survey questionnaires. At an industry level, care has to be taken to ensure that questionnaires are completed by a representative sample of producers, leaning toward neither the best nor the worst, and fully representing any regional differences due to energy use, material sourcing or other factors. The questionnaires cover the full range of inputs and outputs, typically aiming to account for 99% of the mass of a product, 99% of the energy used in its production and any environmentally sensitive flows, even if they fall within the 1% level of inputs.

One area where data access is likely to be difficult is flows from the technosphere. The technosphere is more simply defined as the man-made world. Considered by geologists as secondary resources, these resources are in theory 100% recyclable; however, in a practical sense the primary goal is salvage. For an LCI, these technosphere products (supply chain products) are those that have been produced by man and unfortunately those completing a questionnaire about a process which uses man-made product as a means to an end will be unable to specify how much of a given input they use. Typically, they will not have access to data concerning inputs and outputs for previous production processes of the product. The entity undertaking the LCA must then turn to secondary sources if it does not already have that data from its own previous studies. National databases or data sets that come with LCA-practitioner tools, or that can be readily accessed, are the usual sources for that information. Care must then be taken to ensure that the secondary data source properly reflects regional or national conditions.

Life Cycle Impact Assessment

Inventory analysis is followed by impact assessment. This phase of LCA is aimed at evaluating the significance of potential environmental impacts based on the LCI flow results. Classical life cycle impact assessment (LCIA) consists of the following mandatory elements:

- Selection of impact categories, category indicators, and characterization models;

- The classification stage, where the inventory parameters are sorted and assigned to specific impact categories; and

- Impact measurement, where the categorized LCI flows are characterized, using one of many possible LCIA methodologies, into common equivalence units that are then summed to provide an overall impact category total.

In many LCAs, characterization concludes the LCIA analysis; this is also the last compulsory stage according to ISO 14044:2006. However, in addition to the above mandatory LCIA steps, other optional LCIA elements – normalization, grouping, and weighting – may be conducted depending on the goal and scope of the LCA study. In normalization, the results of the impact categories from the study are usually compared with the total impacts in the region of interest, the U.S. for example. Grouping consists of sorting and possibly ranking the impact categories. During weighting, the different environmental impacts are weighted relative to each other so that they can then be summed to get a single number for the total environmental impact. ISO 14044:2006 generally advises against weighting, stating that "weighting, shall not be used in LCA studies intended to be used in comparative assertions intended to be disclosed to the public". This advice is often ignored, resulting in comparisons that can reflect a high degree of subjectivity as a result of weighting.

Interpretation

Life Cycle Interpretation is a systematic technique to identify, quantify, check, and evaluate information from the results of the life cycle inventory and/or the life cycle impact assessment. The results from the inventory analysis and impact assessment are summarized during the interpretation phase. The outcome of the interpretation phase is a set of conclusions and recommendations for the study. According to ISO 14040:2006, the interpretation should include:

- Identification of significant issues based on the results of the LCI and LCIA phases of an LCA;

- Evaluation of the study considering completeness, sensitivity and consistency checks; and

- Conclusions, limitations and recommendations.

A key purpose of performing life cycle interpretation is to determine the level of confidence in the final results and COMMUNICATE them in a fair, complete, and accurate manner. Interpreting the results of an LCA is not as simple as "3 is better than 2, therefore Alternative A is the best choice"! Interpreting the results of an LCA starts with understanding the accuracy of the results, and ensuring they meet the goal of the study. This is accomplished by identifying the data elements that contribute significantly to each impact category, evaluating the sensitivity of these significant data elements, assessing the completeness and consistency of the study, and drawing conclusions and recommendations based on a clear understanding of how the LCA was conducted and the results were developed.

Reference Test

More specifically, the best alternative is the one that the LCA shows to have the least cradle-to-grave environmental negative impact on land, sea, and air resources.

LCA Uses

Based on a survey of LCA practitioners carried out in 2006 LCA is mostly used to support business strategy (18%) and R&D (18%), as input to product or process design (15%), in education (13%) and for labeling or product declarations (11%). LCA will be continuously integrated into the built environment as tools such as the European ENSLIC Building project guidelines for buildings or developed and implemented, which provide practitioners guidance on methods to implement LCI data into the planning and design process.

Major corporations all over the world are either undertaking LCA in house or commissioning studies, while governments support the development of national databases to support LCA. Of particular note is the growing use of LCA for ISO Type III labels called Environmental Product Declarations, defined as "quantified environmental data for a product with pre-set categories of parameters based on the ISO 14040 series of standards, but not excluding additional environmental information". These third-party certified LCA-based labels provide an increasingly important basis for assessing the relative environmental merits of competing products. Third-party certification plays a major role in today's industry. Independent certification can show a company's dedication to safer and environmental friendlier products to customers and NGOs.

LCA also has major roles in environmental impact assessment, integrated waste management and pollution studies.

Data Analysis

A life cycle analysis is only as valid as its data; therefore, it is crucial that data used for the completion of a life cycle analysis are accurate and current. When comparing different life cycle analyses with one another, it is crucial that equivalent data are available for both products or processes in question. If one product has a much higher availability of data, it cannot be justly compared to another product which has less detailed data.

There are two basic types of LCA data – unit process data and environmental input-output data (EIO), where the latter is based on national economic input-output data. Unit process data are derived from direct surveys of companies or plants producing the product of interest, carried out at a unit process level defined by the system boundaries for the study.

Data validity is an ongoing concern for life cycle analyses. Due to globalization and the rapid pace of research and development, new materials and manufacturing methods are continually being introduced to the market. This makes it both very important and very difficult to use up-to-date information when performing an LCA. If an LCA's conclusions are to be valid, the data must be recent; however, the data-gathering process takes time. If a product and its related processes have not undergone significant revisions since the last LCA data was collected, data validity is not a problem. However, consumer electronics such as cell phones can be redesigned as often as every 9 to 12 months, creating a need for ongoing data collection.

The life cycle considered usually consists of a number of stages including: materials extraction, processing and manufacturing, product use, and product disposal. If the most environmentally harmful of these stages can be determined, then impact on the environment can be efficiently reduced by focusing on making changes for that particular phase. For example, the most energy-intensive life phase of an airplane or car is during use due to fuel consumption. One of the most effective ways to increase fuel efficiency is to decrease vehicle weight, and thus, car and airplane manufacturers can decrease environmental impact in a significant way by replacing heavier materials with lighter ones such as aluminium or carbon fiber-reinforced elements. The reduction during the use phase should be more than enough to balance additional raw material or manufacturing cost.

VARIANTS

Cradle-to-grave

Cradle-to-grave is the full Life Cycle Assessment from resource extraction ('cradle') to use phase and disposal phase ('grave'). For example, trees produce paper, which can be recycled into low-energy production cellulose (fiberised paper) insulation, then used as an energy-saving device in the ceiling of a home for 40 years, saving 2,000 times thefossil-fuel energy used in its production. After 40 years the cellulose fibers are replaced and the old fibers are disposed of, possibly incinerated. All inputs and outputs are considered for all the phases of the life cycle.

Cradle-to-gate

Cradle-to-gate is an assessment of a *partial* product life cycle from resource extraction (*cradle*) to the factory gate (*i.e.*, before it is transported to the consumer). The use phase and disposal phase of the product are omitted in this case. Cradle-to-gate assessments are sometimes the basis for environmental product declarations (EPD) termed business-to-business EDPs. One of the significant uses of the cradle-to-gate approach compiles the life cycle inventory (LCI) using cradle-to-gate. This allows the LCA to collect all of the impacts leading up to resources being purchased by the facility. They can then add the steps involved in their transport to plant and manufacture process to more easily produce their own cradle-to-gate values for their products.

Cradle-to-cradle or Closed Loop Production

Cradle-to-cradle is a specific kind of cradle-to-grave assessment, where the end-of-life disposal step for the product is a recycling process. It is a method used to minimize the environmental impact of products by employing sustainable production, operation, and disposal practices and aims to incorporate social responsibility into product development. From the recycling process originate new, identical products (*e.g.*, asphalt pavement from discarded asphalt pavement, glass bottles from collected glass bottles), or different products (*e.g.*, glass wool insulation from collected glass bottles).

Allocation of burden for products in open loop production systems presents considerable challenges for LCA. Various methods, such as the avoided burden approach have been proposed to deal with the issues involved.

Gate-to-gate

Gate-to-gate is a partial LCA looking at only one value-added process in the entire production chain. Gate-to-gate modules may also later be linked in their appropriate production chain to form a complete cradle-to-gate evaluation.

Well-to-wheel

Well-to-wheel is the specific LCA used for transport fuels and vehicles. The analysis is often broken down into stages entitled "well-to-station", or "well-to-tank", and "station-to-wheel" or "tank-to-wheel", or "plug-to-wheel". The first stage, which incorporates the feedstock or fuel production and processing and fuel delivery or energy transmission, and is called the "upstream" stage, while the stage that deals with vehicle operation itself is sometimes called the "downstream" stage. The well-to-wheel analysis is commonly used to assess total energy consumption, or the energy conversion efficiency and emissions impact of marine vessels, aircraft and motor vehicles, including their carbon footprint, and the fuels used in each of these transport modes.

The well-to-wheel variant has a significant input on a model developed by the Argonne National Laboratory. The Greenhouse gases, Regulated Emissions, and Energy use in Transportation (GREET) model was developed to evaluate the impacts of new fuels and vehicle technologies. The model evaluates the impacts of fuel use using a well-to-wheel evaluation while a traditional cradle-to-grave approach is used to determine the impacts from the vehicle itself. The model reports energy use, greenhouse gas emissions, and six additional pollutants: volatile organic compounds (VOCs), carbon monoxide (CO), nitrogen oxide (NO_x), particulate matter with size smaller than 10 micrometre (PM10), particulate matter with size smaller than 2.5 micrometre (PM2.5), and sulfur oxides (SO_x).

Economic Input–output Life Cycle Assessment

Economic input–output LCA (EIOLCA) involves use of aggregate sector-level data on how much environmental impact can be attributed to each sector of the economy and how much each sector purchases from other sectors. Such analysis can account for long chains (for example, building an automobile requires energy, but producing energy requires vehicles, and building those vehicles requires energy, *etc.*), which somewhat alleviates the scoping problem of process LCA; however, EIOLCA relies on sector-level averages that may or may not be representative of the specific subset of the sector relevant to a particular product and therefore is not suitable for evaluating the environmental impacts of products. Additionally the translation of economic quantities into environmental impacts is not validated.

Ecologically Based LCA

While a conventional LCA uses many of the same approaches and strategies as an Eco-LCA, the latter considers a much broader range of ecological impacts. It was designed to provide a guide to wise management of human activities by understanding the direct and indirect impacts on ecological resources and surrounding ecosystems. Developed by Ohio State University Center for resilience, Eco-LCA is a methodology that quantitatively takes into account regulating and supporting services during the life cycle of economic goods and products. In this approach services are categorized in four main groups: supporting, regulating, provisioning and cultural services.

Life Cycle Energy Analysis

Life cycle energy analysis (LCEA) is an approach in which all energy inputs to a product are accounted for, not only direct energy inputs during manufacture, but also all energy inputs needed to produce components, materials and services needed for the manufacturing process. An earlier term for the approach was *energy analysis*.

With LCEA, the *total life cycle energy input* is established.

Energy Production

It is recognized that much energy is lost in the production of energy commodities themselves, such as nuclear energy, photovoltaic electricity or high-quality petroleum products.*Net energy content* is the energy content of the product minus energy input used during extraction and conversion, directly or indirectly. A controversial early result of LCEA claimed that manufacturing solar cells requires more energy than can be recovered in using the solar cell. The result was refuted. Another new concept that flows from life cycle assessments is Energy Cannibalism. Energy Cannibalism refers to an effect where rapid growth of an entire energy-intensive industry creates a need for energy that uses (or cannibalizes) the energy of existing power plants. Thus during rapid growth the industry as a whole produces no energy because new energy is used to fuel the embodied energy of future power plants. Work has been undertaken in the UK to determine the life cycle energy (alongside full LCA) impacts of a number of renewable technologies.

Energy Recovery

If materials are incinerated during the disposal process, the energy released during burning can be harnessed and used for electricity production. This provides a low-impact energy source, especially when compared with coal and natural gas While incineration produces more greenhouse gas emissions than landfilling, the waste plants are well-fitted with filters to minimize this negative impact. A recent study comparing energy consumption and greenhouse gas emissions from landfilling (without energy recovery) against incineration (with energy recovery) found incineration to be superior in all cases except for when landfill gas is recovered for electricity production.

Criticism

A criticism of LCEA is that it attempts to eliminate monetary cost analysis, that is replace the currency by which economic decisions are made with an energy currency. It has also been argued that energy efficiency is only one consideration in deciding which alternative process to employ, and that it should not be elevated to the only criterion for determining environmental acceptability; for example, simple energy analysis does not take into account the renewability of energy flows or the toxicity of waste products; however the life cycle assessment does help companies become more familiar with environmental properties and improve their environmental system. Incorporating Dynamic LCAs of renewable energy technologies (using sensitivity analyses to project future improvements in renewable systems and their share of the power grid) may help mitigate this criticism.

In recent years, the literature on life cycle assessment of energy technology has begun to reflect the interactions between the current electrical grid and future energy technology. Some papers have focused on energy life cycle, while others have focused on carbon dioxide (CO_2) and other greenhouse gases. The essential critique given by these sources is that when considering energy technology, the growing nature of the power grid must be taken into consideration. If this is not done, a given class of energy technology may emit more CO_2 over its lifetime than it mitigates.

A problem the energy analysis method cannot resolve is that different energy forms (heat, electricity, chemical energy *etc.*) have different quality and value even in natural sciences, as a consequence of the two main laws of thermodynamics. A thermodynamic measure of the quality of energy is exergy. According to the first law of thermodynamics, all energy inputs should be accounted with equal weight, whereas by the second law diverse energy forms should be accounted by different values.

The conflict is resolved in one of these ways:

- value difference between energy inputs is ignored,
- a value ratio is arbitrarily assigned (*e.g.*, a joule of electricity is 2.6 times more valuable than a joule of heat or fuel input),
- the analysis is supplemented by economic (monetary) cost analysis,
- exergy instead of energy can be the metric used for the life cycle analysis.

Critiques

Life cycle assessment is a powerful tool for analyzing commensurable aspects of quantifiable systems. Not every factor, however, can be reduced to a number and inserted into a model. Rigid system boundaries make accounting for changes in the system difficult. This is sometimes referred to as the boundary critique to systems thinking. The accuracy and availability of data can also contribute to inaccuracy. For instance, data from generic processes may be based on averages, unrepresentative sampling, or outdated results. Additionally, social implications of products are generally lacking in LCAs. Comparative life-cycle analysis is often used to determine a better process or product to use. However, because of aspects like differing system boundaries, different statistical information, different product uses, *etc.*, these studies can easily be swayed in favor of one product or process over another in one study and the opposite in another study based on varying parameters and different available data. There are guidelines to help reduce such conflicts in results but the method still provides a lot of room for the researcher to decide what is important, how the product is typically manufactured, and how it is typically used.

An in-depth review of 13 LCA studies of wood and paper products found a lack of consistency in the methods and assumptions used to track carbon during the product life cycle. A wide variety of methods and assumptions were used, leading to different and potentially contrary conclusions – particularly with regard to carbon sequestration and methane generation in landfills and with carbon accounting during forest growth and product use.

Streamline LCA

This process includes three steps. First, a proper method should be selected to combine adequate accuracy with acceptable cost burden in order to guide decision making. Actually, in LCA process, besides streamline LCA, Eco-screening and complete LCA are usually considered as well. However, the former one only could provide limited details and the latter one with more detailed information is more expensive. Second, single measure of stress should be selected. Typical LCA output includes resource consumption, energy consumption, water consumption, emission of CO_2, toxic residues and so on. One of these output is used as the main factor to measure in streamline LCA. Energy consumption and CO_2 emission are often regarded as "practical indicators". Last, stress selected in step 2 is used as standard to assess phase of life separately and identify the most damaging phase. For instance, for a family car, energy consumption could be used as the single stress factor to assess each phase of life. The result shows that the most energy intensive phase for a family car is usage stage.

GOAL & SCOPE DEFINITION

As with other assessments, the first step involves clarifying the purpose and extent of the LCA. This entails formally determining the functional unit, impacts of interest, and system boundary — elements from our "First Choice".

While LCA "light" approaches have been described above, a "full" LCA includes actual primary environmental impact data gathered once the product's full lifecycle has been determined. Such detailed LCAs take, on average, three months and cost $10,000-$60,000, and are only possible to complete once the product is in use and has gone through all stages of its life cycle. This increased accuracy is worth it for benchmarking or external reporting (such as green marketing) purposes.

Inventory Analysis

The next phase entails creating a list of all of the components of the products life cycle that fall within the defined system boundary. It has three major steps:

1. Construct a process flowchart that shows the following:
    *   Raw materials
    *   Mfg processes
    *   Transports
    *   Uses
    *   Waste management
2. Collect data for:
    *   Material inputs
    *   Products and byproducts
    *   Solid waste, air and water emissions
3. Calculate the amounts of each in relation to the functional unit

Essentially, this is the process flow diagram — with detailed mass and energy values attached — that Tom and Priscilla sketched out. The resulting Life Cycle Inventory (LCI) provides a breakdown of all of the energy and materials involved in a product's system at a level of detail that provides a basis for evaluation.

Impact Assessment

Once a detailed LCI is created, environmental impacts can be ascribed to its parts, and if desired to the whole system. There are four steps to the Life Cycle Impact Assessment (LCIA) process, the first two of which are considered mandatory, while the last two are optional.

1. *Classification:* Classification involves assigning specific environmental impacts to each component of the LCI. It is here where decisions made during the scope and goal phase about what environmental impact categories are of interest come into play.

2. *Characterization:* Once the impact categories have been identified, conversion factors – generally known as characterization or equivalency factors – use formulas to convert the LCI results into directly comparable impact indicators.

3. *Normalization (optional):* Some practitioners choose to normalize the impact assessment by scaling the data by a reference factor, such as the region's per capita environmental burden. This helps to clarify the relative impact of a substance in a given context. For instance, if global warming contributions are already high in the context in which the product is being assessed, a reference factor would normalize whatever the product's global warming contributions are in order to clarify its relative impacts.

4. *Weighting (optional):* The pros and cons of weighting were described in the Measurements section above.

Interpretation

Although listed fourth, life cycle interpretation actually occurs throughout the whole LCA. It involves the ongoing process of clarifying, quantifying, checking, and evaluating the information used by, and resulting from, the life cycle inventory (LCI) and impact assessment (LCIA) phases. The standard that covers the LCA process, ISO 14044, gives two main objectives:

1. Analyze results, reach conclusions, explain limitations, and provide recommendations based on the findings of the preceding phases of the LCA, and to report the results of the life cycle interpretation in a transparent manner.

2. Provide a readily understandable, complete, and consistent presentation of the results of an LCA study, in accordance with the goal and scope of the study.

To achieve these objectives, the ISO standard states that interpretation should cover at least three major elements.

1. Identification of the significant issues based on the LCI and LCIA. Which life cycle stages or components stand out as major contributors to overall impact? What are the anomalies?

2. Evaluation which considers completeness, sensitivity, and consistency checks. Is all the information needed for interpretation present in the LCI and LCIA? How reliable is the information related to any identified significant issues? How much do changes in such factors influence the overall results? Are all

of the assumptions, data, characterization factors, *etc.* that were used in the assessment consistent internally and with the overall goal and scope of the LCA?

It is very important to note that no matter how carefully assembled, analyzed, assessed, and measured, LCAs are never the "real" answer. They require interpretation, which is turn requires transparency and judgment. The data sources, assumptions, and all other relevant information needs to be transparent to decision makers so that they can understand the full context of the results of the life cycle inventory assessment. Deciding among design options is not as easy as just comparing LCIA numbers, whether single- or multi-factor, weighted or not. LCIA results can be a source of insights, but do not stand alone in guiding product development choices. Engineers will need to take them in the context of the other attributes they are trying to optimize, including cost, manufacturability, performance, and so on. In addition, there are myriad other factors guiding product development decisions not covered by LCAs, including social impacts and acceptance, pricing, political agendas, and regulations.

PHASES OF LIFE CYCLE ASSESSMENT (LCA)

The Phases of Life Cycle Assessment

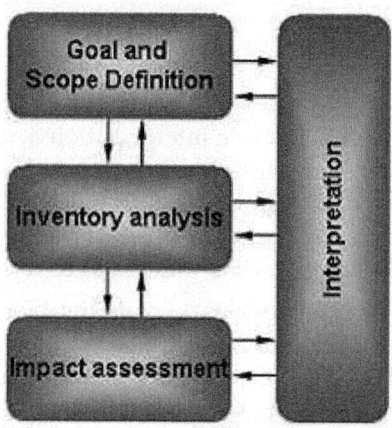

1. Goal and Scope Definition, the product(s) or service(s) to be assessed are defined, a functional basis for comparison is chosen and the required level of detail is defined;
2. Inventory Analysis of extractions and emissions, the energy and raw materials used, and emissions to the atmosphere, water and land, are quantified for each process, then combined in the process flow chart and related to the functional basis;
3. Impact Assessment, the effects of the resource use and emissions generated are grouped and quantified into a limited number of impact categories which may then be weighted for importance;

4.   Interpretation, the results are reported in the most informative way possible and the need and opportunities to reduce the impact of the product(s) or service(s) on the environment are systematically evaluated.

PHASE 1: GOAL AND SCOPE DEFINITION

Goal and scope definition

The first part of an LCA study consists of defining the goal of the study and its scope. The goal of the study should include a statement of the reason for carrying out the study as well as the intended application of the results and the intended audience. In the scope of an LCA the following items shall be considered and described:

- The function of the product system.
- The functional unit.
- The system boundaries.
- Allocation procedures.
- Type of impact assessment methodology and interpretation to be performed.
- Data requirements.
- Assumptions and limitations.
- Data quality requirements.
- Type of critical review, if any.
- Type and format of the report required for the study.

The scope should describe the depth of the study and show that the purpose can be fulfilled with the actual extent of the limitations.

Functional Unit

The functional unit is a key element of LCA which has to be clearly defined. The functional unit is a measure of the function of the studied system and it provides a reference to which the inputs and outputs can be related. This enables comparison of two essential different systems. For example, the functional unit for a paint system may be defined as the unit surface protected for 10 years. A comparison of the environmental impact of two different paint systems with the same functional unit is therefore possible.

System Boundaries

The system boundaries determine which unit processes to be included in the LCA study. Defining system boundaries is partly based on a subjective choice, made during the scope phase when the boundaries are initially set. The following boundaries can be considered:

- Boundaries between the technological system and nature. A life cycle usually begins at the extraction point of raw materials and energy carriers from nature. Final stages normally include waste generation and/or heat production.

- Geographical area. Geography plays a crucial role in most LCA studies, *e.g.* infrastructures, such as electricity production, waste management and transport systems, vary from one region to another. Moreover, ecosystems sensitivity to environmental impacts differs regionally too.

- Time horizon. Boundaries must be set not only in space, but also in time. Basically LCAs are carried out to evaluate present impacts and predict future scenarios. Limitations to time boundaries are given by technologies involved, pollutants lifespan, *etc.*

- Boundaries between the current life cycle and related life cycles of other technical systems. Most activities are interrelated, and therefore must be isolated from each other for further study. For example production of capital goods, economic feasibility of new and moreenvironmentally friendly processes can be evaluated in comparison with currently used technology.

Data Quality Requirements

Reliability of the results from LCA studies strongly depends on the extent to which data quality requirements are met. The following parameters should be taken into account:

- Time-related coverage.
- Geographical coverage.
- Technology coverage.
- Precision, completeness and representativeness of the data.
- Consistency and reproducibility of the methods used throughout the data collection.
- Uncertainty of the information and data gaps.

Reusability of data is also highly dependent on sufficient data documentation. One example of a format for sufficient environmental data documentation is the LCI data documentation software SPINE@CPM Data Tool and the LCI database SPINE@CPM Database. These where developed within the CPM collaboration, in order to enable an effective and efficient handling of environmental information.

The ISO/TS 14048 data documentation format, described in reports found under reports on this site, is another format for transparent, reusable documentation focusing on data quality.

Inventory Analysis (LCI)

LCI comprises all stages dealing with data retrieval and management. The data collection forms must be properly designed for optimal collection. Subsequently data are validated and related to the functional unit in order to allow the aggregation of results. A very sensitive step in this calculation process is the allocation of flows *e.g.* releases to air, water and land. Most of the existing technical systems yield more than one product. Therefore, materials and energy flows regarding the process as a whole, as well as environmental releases must often be allocated to the different products. This is recommended to be made according to a given procedure:

- Wherever possible, allocation should be avoided.

- Where allocation is not avoidable, inputs and outputs should be partitioned between its different functions or products in a way that reflects the underlying physical relationships between them.

- If the latter is not possible, allocation should be carried out based on other existing relationships (*e.g.* in proportion to the economic value of products).

The data collection is the most resource consuming part of the LCA. Reuse of data from other studies can simplify the work but this must be made with great care so that the data is representative. The quality aspect is therefore also crucial.

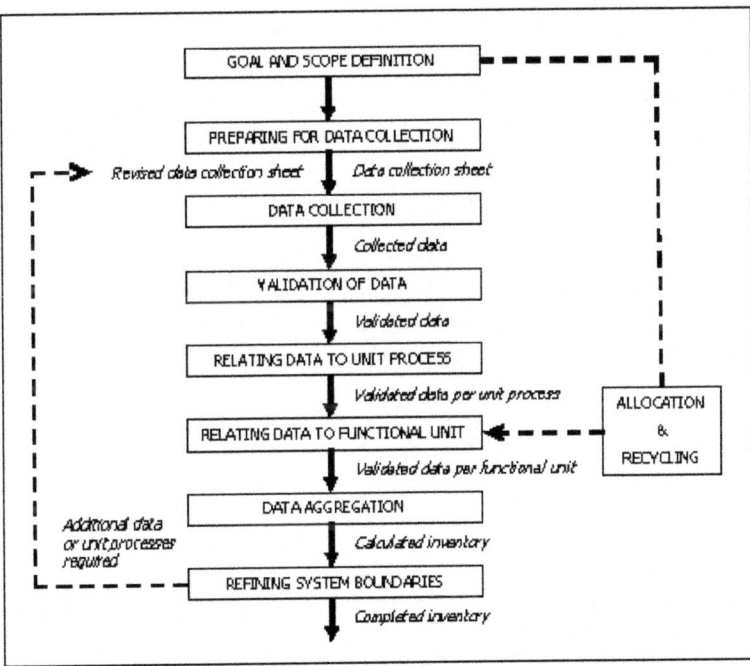

Impact Assessment (LCIA)

LCIA aims to evaluate the significance of potential environmental impacts using the results coming out from the LCI phase. The ISO14040 suggests that this phase of an LCA is divided into the following steps:

Mandatory elements:

- Selection of impact categories, category indicators and characterization models.
- Classification, *i.e.* assignment of individual inventory parameters to impact categories, *e.g.* CO_2 is assigned to Global Warming. Common impact categories are Global Warming, Ozone Depletion, Photooxidant Formation, Acidification and Eutrophication.
- Characterization, *i.e.* conversion of LCI results to common units within each impact category, so that results can be aggregated into category indicator results.

Optional elements:

- Normalization. The magnitude of the category indicator results is calculated relatively to reference information, *e.g.* and old products constitutes baseline when assigning a new product.
- Weighting. Indicator results coming from the different impact categories are converted to a common unit by using factors based on value-choices.
- Grouping. The impact categories are assigned into one or more groups sorted after geographic relevance, company priorities *etc.*

The methods that are usually used for LCIA are *e.g.* EPS (Environmental Priority Strategies), ECO (Ecological scarcity) and ET (Environmental Theme).

Interpretation

The aim of the interpretation phase is to reach conclusions and recommendations in accordance with the defined goal and scope of the study. Results from the LCI and LCIA are combined together and reported in order to give a complete and unbiased account of the study. The interpretation is to be made iteratively with the other phases.

The life cycle interpretation of an LCA or an LCI comprises three main elements:

- Identification of the significant issues based on the results of the LCI and LCIA phases of a LCA.
- Evaluation of results, which considers completeness, sensitivity and consistency checks.
- Conclusions and recommendations.

In ISO 14040 standard it is recommended that a critical review should be performed. In addition it is stated that a critical review must have been conducted in order to disclose the results in public.

PHASE 2: INVENTORY ANALYSIS

Inventory Analysis (ISO 14041)

This phase is the most work intensive and time consuming of all the phases in an LCA, this is mainly because of the data collection. The phase includes:

- Data collection, quantitative and qualitative data for every unit process in the process system
- Relate data to the functional unit
- Make continuous validation of the data
- Use allocation or system expansion where necessary
- Refine system boundaries

Data Collection

The data collection must be performed according to the functional unit and system boundaries. The data that should be used is for all the important processes within the system boundaries.

The data should include all inputs and outputs from the processes. Inputs are for example use of energy, water, materials *etc.* Outputs are the products and co-products. Emissions can be divided into four categories: air, water, soil and solid waste depending on what the emissions affect.

The data collection can be less time consuming if good databases are available and if customers and suppliers are willing to help gather data. A lot of databases with LCA data exist and they can normally be bought together with the LCA software. Data on transport, extraction of raw materials, processing of materials, production of normally used products like plastic; cardboard *etc.* and disposal can normally be found in an LCA-database. Data can also be collected through national statistics as done in the LCAfood database for the farming processes.

Data from the databases can be used for processes that are not product specific, like general data on the production of electricity, coal or packaging. For example in the LCA on milk from the store specific data is needed on the farming while general data from the databases can be used for the production of electricity and fertilizers.

System Expansion and Allocation

In many processes more than one product is produced (joint production), in such cases it is necessary to divide the environmental impacts from the process between the products. It is not straightforward to divide environmental impacts between the product and the co-product, but with help from allocation or system expansion it can be done. The choice between the two methods can have huge impacts on the result of the LCA. The ISO 14040 –series suggest using system expansion whenever possible and where it is not possible to use system expansion allocation can be used instead.

Allocation of environmental impacts between the product and co-products can for example be performed from an economic or weight point of view. Allocation is a division of the environmental impacts according to how much the products costs/weights, *e.g.* if a product costs 60dkr and the co-product costs 40dkr, then the product is said to give 60% of the environmental impacts and the co-product 40% of the environmental impacts. The weight allocation is performed in the same way, just with weight data.

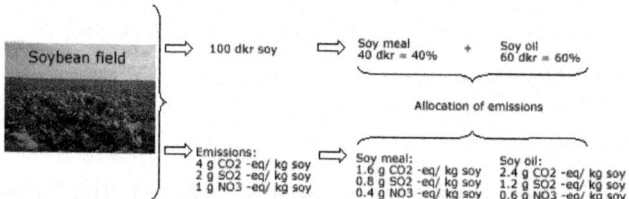

Fig. : Example of allocation. eq. = equivalents.

In system expansion co-products are considered alternatives to other products on the global market as shown in the following example.

At a dairy farm milk is the main product and cattle meat is produced as a co-product. In such a production it is not possible to allocate precisely what feed use, land use, emissions *etc.* are related to for example milk vs. meat and therefor system expansion must be used. The cattle meat production replaces cattle meat from Argentina or pork meat on the market depending on the price of the Danish cattle meat, which means that there is an avoided production of Argentinean cattle meat or Danish pork and thereby a negative contribution to the environmental impact from milks lifecycle. In this example the system is then expanded to include the system of processes, which are involved in Argentinean cattle meat or the pork meat.

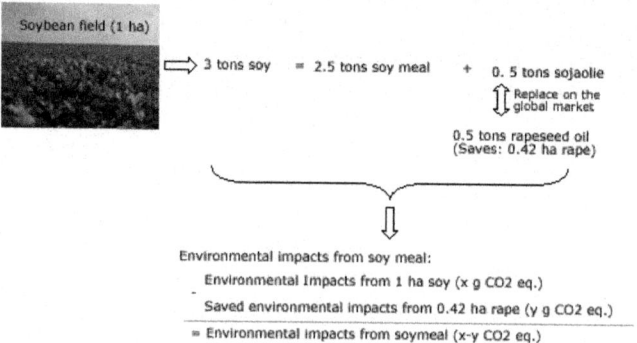

Fig. : Example of system expansion. eq. = equivalents.

Identification of the marginal product can be made trough a marginal line of thinking, where the avoided product is a marginal product, which is the product that will be used more of if the production of the co-product decreases. Thorough the crediting the main product forms a system of products and avoided products.

Another example where system expansion is necessary is in flour. Grains are processed into flour and bran during the milling process. The environmental emissions associated with flour are determined by summarizing the total emissions associated with the milling process and subtracting the total emissions associated with the product that the marginal bran displaces in the market. Marginal bran is used for animal feed and is assumed to displace spring barley. Hence, the system of processes involved in analysis of flour is expanded to include the system of processes which are involved in spring barley production and the total emissions associated with flour is determined by subtracting the emissions associated with saved spring barley production from the emissions associated with flour and bran production. "

The two methods for dividing the environmental impacts between the main product and co-product (allocation and system expansion) is part of two methods for making LCAs. Allocation is part of the traditional attributional method. Attributional LCA seeks only to cut the piece of the global environmental impact related to the product. The goal is to describe the environmental relevant physical flow. Average suppliers data are used. System expansion is part of the consequential LCA method that seeks to capture change in environmental impact as a consequence of a certain activity and there by generate information on consequences of actions. Marginal data is used.

In the LCAfood database the consequential LCA method is used, this method is the most precise to use because it reflects reality and the market forces. The consequential method includes all processes that are affected by the production of a product *e.g.* the production of milk, where the production of meat is affected and there by included in the LCA as an avoided production. In attributional LCA the allocation is made based on the price or weight of the product and thereby the environmental impacts are not divided between the products according to the environmental reality but according to more or less arbitrary partitioning coefficients.

PHASE 3: IMPACT ASSESSMENT

In this phase the collected data are processed and the actual result of the LCA is given. The emissions of a product are divided into different environmental impact categories. The result can be presented in different ways, for example for the entire lifecycle or for single parts of the lifecycle but always within theenvironmental impact categories. An overview of phase 3 is given in figure below followed by a description of the different parts of the figure.

The figure shows on the left some of the processes within the lifecycle of one kilo cheese and the emissions from these processes. On the right side the environmental impact categories that the emissions contribute to can be seen. In the top right corner of the figure the co-products from the production of cheese are seen, these products will result in a system expansion of the life cycle.

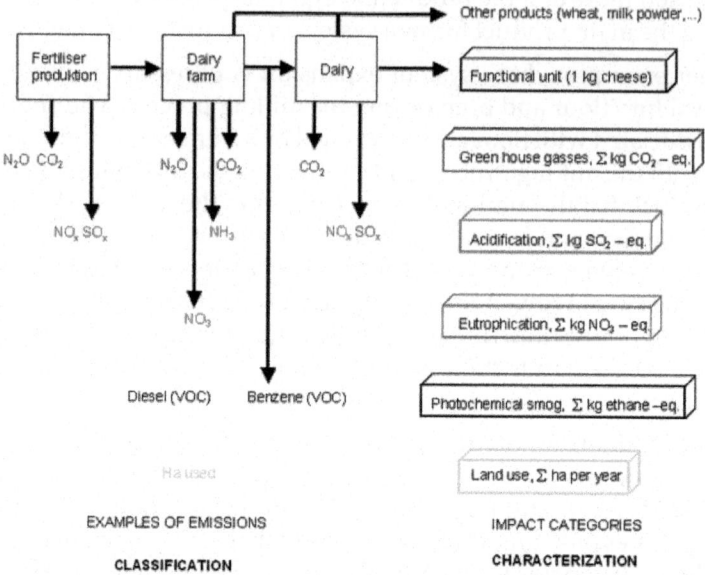

Fig. : Classification and characterization.

The environmental categories are described in the following to give an overview of how each category affect the environment and which human activities contributes to this category.

Environmental impact categories:

- Global warming is a warming of the atmosphere, which causes climate changes. Some of the biggest human contributors to global warming are the combustion of fossil fuels like oil, coal and natural gas. Global warming potential are in the LCAFood database presented in g CO_2-equivalents.

- Acidification is caused by acids and compounds which can be converted into acids that contributes to death of fish and forests, damage on buildings *etc.* The most significant man made sources of acidification are combustion processes in electricity and heating production, and transport. Acidification potentials are in the LCAFood database presented in g SO_2- equivalents.

- Eutrophication also called nutrient enrichment causes algal bloom in inlets and springs causing oxygen depletion and death of fish. Emissions of nitrogen to the aquatic environment, especially fertilizers from agriculture contribute to eutrophication. Also oxides of nitrogen from combustion processes are of significance. Eutrophication potentials are in the LCAFood database presented in g NO_3- equivalents.

- Photochemical smog formation occurs when Volatile Organic Compounds (VOC's) are released in the atmosphere and oxidized in the presence of oxides of nitrogen (NO_x). The most significant VOC's emissions from unburnt petrol and diesel and the use of organic solvents, like paints. Photochemical smog attacks organic compounds in plants, animals and materials exposed to air, causing problems in the respiratory tract in humans. For agriculture it causes a reduction in yield. Photochemical smog formation potentials are in the LCAFood database presented in g ethane equivalents.

- Ozone depletion: Stratospheric ozone is broken down as a consequence of man-made emissions of halocarbons (CFC's, HCFC's, haloes, chlorine, bromine *etc.*). The ozone content of the stratosphere is therefore decreasing and thinning of ozone layer, often referred to as the ozone hole. The consequences are increased frequency of skin cancer in humans and damage to the plants.

- Land use: Area of land used in the production of a product, for example agricultural land. Land use is presented in hectare year (ha*yr) or in m^2*yr.

Toxic impact categories also exist; these are not described here. It is difficult to get valid result within the toxic categories because it is difficult to get useful and precise data on the processes that contribute to toxic categories. This can for example be on production of chemicals.

Calculations in a LCA (Five Steps)

In phase 3 the result of a LCA is given through five calculation steps, these are:

1. Classification: Review of the inventory deciding to which impacts categories the emissions contribute, *e.g.* global warming.

2. Characterization: Calculation of the emission's potential environmental impact is done by using equivalence factors specific for each impact category. This way different substances emitted can be transformed into the same unit. For the global warming the reference is for example normally CO_2.

3. Normalization (PE): In normalisatin the environmental impacts are seen in relation to an average danish persons contribution to the environmental effect during a year. This means that normalisation is a relative magnitude of the potential impacts and resource consumptions. The environmental impacts from the characterization are compared with an impact, which is common for all impacts categories and of which the consequences for the environment, resources and working environment are known. In the EDIP method, which is used in the LCA food database, the resource consumption and the potential impacts which society imposes on the environment each year are used for the comparison. The unit is impact potential per person per year.

4. Weighting (*e.g.* wPE value choices), the environmental impacts form the normalization is compared to the political reduction goals. The seriousness of the impacts categories relative to one another is made. Weighting ables a comparison of the potentials for the various impacts.

5. Sensitivity analysis, estimation on how much the results can vary. The sensitivity analysis covers the product system model, the processes and the assessment factors.

In the LCAfood database the data has only been classified and characterized, examples of characterized impact potentials from full milks lifecycle. It is up to the user of the LCAfood database to perform the normalization them self. It is in most cases not advisable to weight the result because of many uncertainties with the weighting, caused by the politically set reduction goals.

PHASE 4: INTERPRETATION

Phase 4 is performed through each of the other steps in the LCA to ensure the quality of the LCA. The phase include:

*   Analyze results: Identification of significant issues
*   Evaluation: Completeness, sensitivity and consistency checks (quality check)
*   Reach conclusion and explain limitations
*   Provide recommendations

Analyze Results

In this phase the results from the Life cycle are analyzed according to the decisions made in Phase 1 about scope of the LCA. The environmental impacts from full milks lifecycle are shown, the impacts are divided into some of the major phases in milks life cycle. It can be seen that the biggest contributor in all the impact categories is the farming. Farming also includes the production of fodder, electricity, fertilizers *etc.* used in the farming process.

The LCAfood database meets the demands in the ISO 14040-series, this with the assumption that the system expansion has been performed with the right processes and thereby that the avoided processes are correct. The LCAfood database is the first database using system expansion systematically, which is suggested as first priority in the ISO 14040-series, allocation is second.

Links to Other Relevant Information on LCAs

Wenzel H, Hauschild M and Alting L (2001): Environmental Assessment of Products, Volume 1: Methodology, tools and case studies in product development, ISBN 0-412-80800-5, Institute for Product Development

THERMODYNAMIC INPUT-OUTPUT IN LIFE CYCLE ASSESSMENT (LCA)

Thermodynamic Cycle

A thermodynamic cycle consists of a linked sequence of thermodynamic processes that involve transfer of heat and work into and out of the system, while varying pressure, temperature, and other state variables within the system, and that eventually returns the system to its initial state. In the process of passing through a cycle, the working fluid (system) may convert heat from a warm source into useful work, and dispose of the remaining heat to a cold sink, thereby acting as a heat engine. Conversely, the cycle may be reversed and use work to move heat from a cold source and transfer it to a warm sink thereby acting as a heat pump.

During a closed cycle, the system returns to its original thermodynamic state of temperature and pressure. Process quantities (or path quantities), such as heat and work are process dependent. For a cycle for which the system returns to its initial state the first law of thermodynamics applies:

$$\Delta E = E_{out} - E_{in} = 0$$

The above states that there is no change of the energy of the system over the cycle. E_{in} might be the work and heat input during the cycle and E_{out} would be the work and heat output during the cycle. The first law of thermodynamics also dictates that the net heat input is equal to the net work output over a cycle (we account for heat, Q_{in}, as positive and Q_{out} as negative). The repeating nature of the process path allows for continuous operation, making the cycle an important concept in thermodynamics. Thermodynamic cycles are often represented mathematically as quasistatic processes in the modeling of the workings of an actual device.

Heat and Work

Two primary classes of thermodynamic cycles are power cycles and heat pump cycles. Power cycles are cycles which convert some heat input into a mechanical work output, while heat pump cycles transfer heat from low to high temperatures by using mechanical work as the input. Cycles composed entirely of quasistatic processes can operate as power or heat pump cycles by controlling the process direction. On a pressure-volume (PV) diagram or temperature-entropy diagram, the clockwise and counterclockwisedirections indicate power and heat pump cycles, respectively.

Relationship to Work

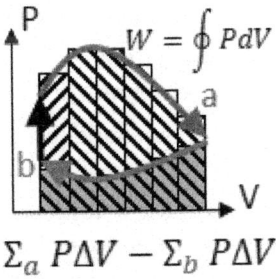

$$\Sigma_a \, P\Delta V - \Sigma_b \, P\Delta V$$

Fig. : The net work equals the area inside because it is (a) the Riemann sum of work done on the substance due to expansion, minus (b) the work done to re-compress.

Because the net variation in state properties during a thermodynamic cycle is zero, it forms a closed loop on a PV diagram. A PV diagram's Y axis shows pressure (P) and X axis shows volume (V). The area enclosed by the loop is the work (W) done by the process:

(1) $W = \oint P \, dV$

This work is equal to the balance of heat (Q) transferred into the system:

(2) $W = Q = Q_{in} - Q_{out}$

Equation (2) makes a cyclic process similar to an isothermal process: even though the internal energy changes during the course of the cyclic process, when the cyclic process finishes the system's energy is the same as the energy it had when the process began.

If the cyclic process moves clockwise around the loop, then W will be positive, and it represents a heat engine. If it moves counterclockwise, then W will be negative, and it represents a heat pump.

Each Point in the Cycle

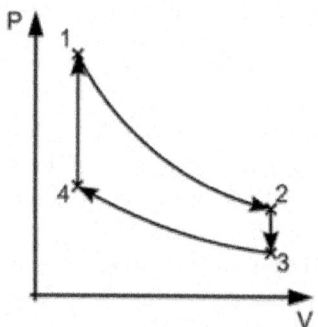

Fig. : Description of each point in the thermodynamic cycles.

Otto Cycle:

1→2: Isentropic Expansion: Constant entropy (s), Decrease in pressure (P), Increase in volume (v), Decrease in temperature (T)

2→3: Isochoric Cooling: Constant volume(v), Decrease in pressure (P), Decrease in entropy (S), Decrease in temperature (T)

3→4: Isentropic Compression: Constant entropy (s), Increase in pressure (P), Decrease in volume (v), Increase in temperature (T)

4→1: Isochoric Heating: Constant volume (v), Increase in pressure (P), Increase in entropy (S), Increase in temperature (T)

A List of Thermodynamic Processes:

Adiabatic : No energy transfer as heat (Q) during that part of the cycle would amount to $\delta Q = 0$. This does not exclude energy transfer as work.

Isothermal : The process is at a constant temperature during that part of the cycle (T=constant, $\delta T = 0$). This does not exclude energy transfer as heat or work.

Isobaric : Pressure in that part of the cycle will remain constant. (P = constant, $\delta P = 0$). This does not exclude energy transfer as heat or work.

Isochoric : The process is constant volume (V = constant, $\delta V = 0$). This does not exclude energy transfer as heat or work.

Isentropic : The process is one of constant entropy (S = constant, $\delta S = 0$). This excludes the transfer of heat but not work.

Power Cycles

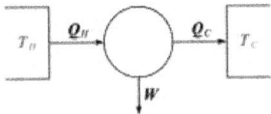

Fig. : Heat engine diagram.

Thermodynamic power cycles are the basis for the operation of heat engines, which supply most of the world's electric power and run the vast majority of motor vehicles. Power cycles can be organized into two categories: real cycles and ideal cycles. Cycles encountered in real world devices (real cycles) are difficult to analyze because of the presence of complicating effects (friction), and the absence of sufficient time for the establishment of equilibrium conditions. For the purpose of analysis and design, idealized models (ideal cycles) are created; these ideal models allow engineers to study the effects of major parameters that dominate the cycle without having to spend significant time working out intricate details present in the real cycle model.

Power cycles can also be divided according to the type of heat engine they seek to model. The most common cycles used to model internal combustion engines are the Otto cycle, which models gasoline engines, and the Diesel cycle, which models diesel engines. Cycles that model external combustion engines include the Brayton cycle, which models gas turbines, the Rankine cycle, which models steam turbines, the Stirling cycle, which models hot air engines, and the Ericsson cycle, which also models hot air engines.

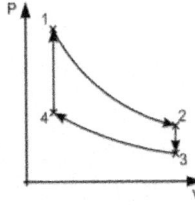

Fig. : The clockwise thermodynamic cycle indicated by the arrows shows that the cycle represents a heat engine. The cycle consists of four states (the point shown by crosses) and four thermodynamic processes (lines).

For example the pressure-volume mechanical work output from the heat engine cycle (net work out), consisting of 4 thermodynamic processes, is:

(3) $W_{net} = W_{1\to2} + W_{2\to3} + W_{3\to4} + W_{4\to1}$

$W_{1\to2} = \int_{V_1}^{V_2} P\,dV$, negative, work done on system

$W_{2\to3} = \int_{V_2}^{V_3} P\,dV$, zero, work if V2 equal V3

$W_{3\to4} = \int_{V_3}^{V_4} P\,dV$, positive, work done by system

$W_{4\to1} = \int_{V_4}^{V_1} P\,dV$, zero, work if V4 equal V1

If no volume change happens in process simplifies to:

(4) $W_{net} = W_{1\to2} + W_{3\to4}$

Heat Pump Cycles

Thermodynamic heat pump cycles are the models for household heat pumps and refrigerators. There is no difference between the two except the purpose of the refrigerator is to cool a very small space while the household heat pump is intended to warm a house. Both work by moving heat from a cold space to a warm space. The most common refrigeration cycle is the vapor compression cycle, which models systems using refrigerants that change phase. The absorption refrigeration cycle is an alternative that absorbs the refrigerant in a liquid solution rather than evaporating it. Gas refrigeration cycles include the reversed Brayton cycle and the Hampson-Linde cycle. Multiple compression and expansion cycles allow gas refrigeration systems to liquify gases.

Modelling Real Systems

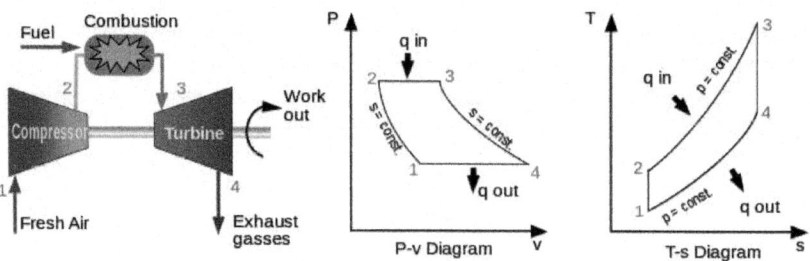

Example of a real system modelled by an idealized process: PV and TS diagrams of a Brayton cycle mapped to actual processes of a gas turbine engine

Thermodynamic cycles may be used to model real devices and systems, typically by making a series of assumptions. Simplifying assumptions are often necessary to reduce the problem to a more manageable form. For example, as shown in the figure, devices such a gas turbine or jet engine can be modeled as a Brayton cycle. The actual device is made up of a series of stages, each of which is itself modeled as an idealized thermodynamic process. Although each stage which acts on the working fluid is a complex real device, they may be modelled as idealized processes which approximate their real behavior. If energy is added by means other than combustion, then a further assumption is that the exhaust gases would be passed from the exhaust to a heat exchanger that would sink the waste heat to the environment and the working gas would be reused at the inlet stage.

The difference between an idealized cycle and actual performance may be significant. For example, the following images illustrate the differences in work output predicted by an ideal Stirling cycle and the actual performance of a Stirling engine:

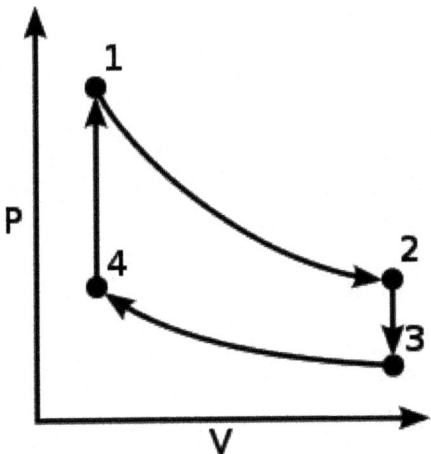

Fig. : Ideal Stirling cycle.

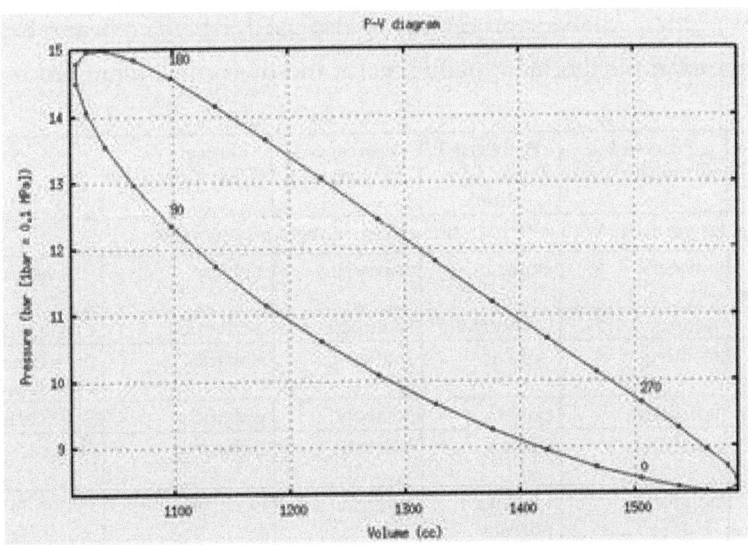

Actual performance.

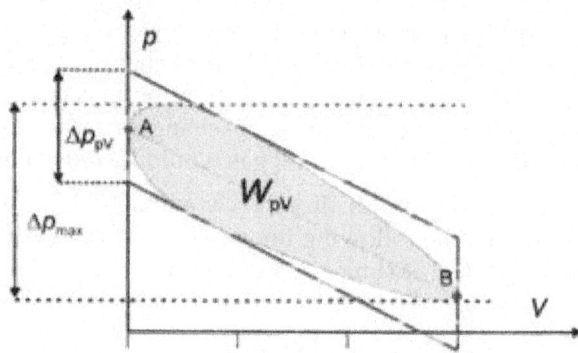

Actual and ideal overlaid, showing difference in work output.

As the net work output for a cycle is represented by the interior of the cycle, there is a significant difference between the predicted work output of the ideal cycle and the actual work output shown by a real engine. It may also be observed that the real individual processes diverge from their idealized counterparts; *e.g.,* isochoric expansion (process 1-2) occurs with some actual volume change.

Well-known Thermodynamic Cycles

In practice, simple idealized thermodynamic cycles are usually made out of four thermodynamic processes. Any thermodynamic processes may be used. However, when idealized cycles are modeled, often processes where one state variable is kept constant are used, such as an isothermal process (constant temperature), isobaric process(constant pressure), isochoric process (constant volume), isentropic process (constant entropy), or an isenthalpic process (constant enthalpy). Often adiabatic processes are also used, where no heat is exchanged.

Some example thermodynamic cycles and their constituent processes are as follows:

Cycle	Process 1-2 (Compression)	Process 2-3 (Heat Addition)	Process 3-4 (Expansion)	Process 4-1 (Heat Rejection)	Notes
Power cycles normally with external combustion - or heat pump cycles:					
Bell Coleman	adiabatic	isobaric	adiabatic	isobaric	A reversed Brayton cycle
Carnot	isentropic	isothermal	isentropic	isothermal	Carnot heat engine
Ericsson	isothermal	isobaric	isothermal	isobaric	the second Ericsson cycle from 1853
Rankine	adiabatic	isobaric	adiabatic	isobaric	Steam engine
Hygro-scopic	adiabatic	isobaric	adiabatic	isobaric	Hygroscopic cycle
Scuderi	adiabatic	variable pressure and volume	adiabatic	isochoric	

Stirling	isothermal	isochoric	isothermal	isochoric	Stirling engine
Stoddard	adiabatic	isobaric	adiabatic	isobaric	
Power cycles normally with internal combustion:					
Brayton	adiabatic	isobaric	adiabatic	isobaric	Jet engines the external combustion version of this cycle is known as first Ericsson cycle from 1833
Diesel	adiabatic	isobaric	adiabatic	isochoric	Diesel engine
Lenoir	isobaric	isochoric	adiabatic		Pulse jets (Note: Process 1-2 accomplishes both the heat rejection and the compression)
Otto	adiabatic	isochoric	adiabatic	isochoric	Gasoline / petrol engines

Ideal Cycle

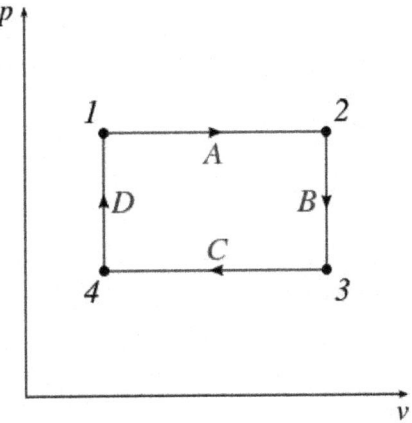

Fig. : An illustration of an ideal cycle heat engine (arrows clockwise).

An ideal cycle is constructed out of:

1. TOP and BOTTOM of the loop: a pair of parallel isobaric processes
2. LEFT and RIGHT of the loop: a pair of parallel isochoric processes

Internal energy of a perfect gas undergoing different portions of a cycle:

Isothermal:

$$\Delta U = RT \ln \frac{V_2}{V_1} - RT \ln \frac{V_2}{V_1} = 0$$ (Note: U of an isothermal process has to equal 0)

Isochoric: $\Delta U = C_v \Delta T - 0 = C_v \Delta T$

Isobaric: $\Delta U = C_p \Delta T - R \Delta T$ (or $P \Delta V$) $= C_v \Delta T$

Carnot Cycle

The Carnot cycle is a cycle composed of the totally reversible processes of isentropic compression and expansion and isothermal heat addition and rejection. The thermal efficiency of a Carnot cycle depends only on the absolute temperatures of the two reservoirs in which heat transfer takes place, and for a power cycle is:

$$\eta = 1 - \frac{T_L}{T_H}$$

where T_L is the lowest cycle temperature and T_H the highest. For Carnot power cycles the coefficient of performance for a heat pump is:

$$COP = 1 + \frac{T_L}{T_H - T_L}$$

and for a refrigerator the coefficient of performance is:

$$COP = \frac{T_L}{T_H - T_L}$$

The second law of thermodynamics limits the efficiency and COP for all cyclic devices to levels at or below the Carnot efficiency. The Stirling cycle and Ericsson cycle are two other reversible cycles that use regeneration to obtain isothermal heat transfer.

Stirling Cycle

A Stirling cycle is like an Otto cycle, except that the adiabats are replaced by isotherms. It is also the same as an Ericsson cycle with the isobaric processes substituted for constant volume processes.

1. TOP and BOTTOM of the loop: a pair of quasi-parallel isothermal processes

2. LEFT and RIGHT sides of the loop: a pair of parallel isochoric processes

Heat flows into the loop through the top isotherm and the left isochore, and some of this heat flows back out through the bottom isotherm and the right isochore, but most of the heat flow is through the pair of isotherms. This makes sense since all the work done by the cycle is done by the pair of isothermal processes, which are described by $Q=W$. This suggests that all the net heat comes in through the top isotherm. In fact, all of the heat which comes in through the left isochore comes out through the right isochore: since the top isotherm is all at the same warmer temperature T_H and the bottom isotherm is all at the same cooler temperature T_C, and since change in energy for an isochore is proportional to change in temperature, then all of the heat coming in through the left isochore is cancelled out exactly by the heat going out the right isochore.

State Functions and Entropy

If Z is a state function then the balance of Z remains unchanged during a cyclic process:

$$\oint dZ = 0.$$

Entropy is a state function and is defined as

$$S = \frac{Q}{T}$$

so that

$$\Delta S = \frac{\Delta Q}{T},$$

then it is clear that for any cyclic process,

$$\oint dS = \oint \frac{dQ}{T} = 0$$

meaning that the net entropy change over a cycle is 0.

ECOLOGICALLY-BASED LIFE CYCLE ASSESSMENT

Innovation is essential for genuine progress in sustainable development. But credible tools are needed to verify the "sustainability" of new products, processes, and technologies. Specifically, companies need to understand the energy, emissions, resource use, and cost trade-offs over the full life cycle of a product or service, from "cradle to cradle". The scope of analysis may include resource extraction, processing, manufacturing, logistics, service, remanufacturing, and end-of-life resource recovery.

Conventional life cycle assessment (LCA), based on ISO 14000 guidelines, is costly and time-intensivedue to the need for accurate, process-specific data. Also, conventional LCA requires setting a boundary that may omit important processes. Alternative approaches have emerged that are more comprehensive, morestreamlined, and less fine-grain than conventional LCA. These macro-scale methods use aggregate input-output data to model the entire economy from a top-down perspective, and provide a useful complement to detailed, bottom-up LCA. In particular, since streamlined LCA requires only basic data about resource inputs, it is helpful in assessing new products when emissions data are not yet available.

Another shortcoming of conventional LCA is lack of attention to natural capital. For example, use of agricultural wastes as a renewable energy source (*e.g.*, for biofuels) can hurt agricultural productivity by reducing the resilience of soil ecosystems. Ecosystem products and services are the foundation of our economy, but are excluded from typical energy and emissions accounting. Ecological product flows include sand, wood, grass, metals, and minerals, while ecological services include water, wind, tides, soil, and pollination. An understanding of threats to these resources is essential for sustainable development.

After years of investigation, Ohio State scientists have devised a rigorous method to quantify the lifecycle resource consumption of industrial products and processes, including "embedded" natural capital. Simply put, this approach extends LCA back to the cradle, providing a full accounting of how economic activities utilize ecosystem products and services. Eco-LCA™ is now available in the form of a webbased softwarepackage, the first tool that enables macro-scale LCA including natural capital. Results are available for 488 industrial sectors of the U.S. economy, and the tool has been applied in a variety of industries, ranging from electric power to consumer products. Eco-LCA™ is capable of analyzing systems at any scale, from a unit process to a supply chain to the entire economy. For example, the figure below illustrates the use of Eco-LCA™ to assess of the relative ecological impacts of consumer expenditures for an average U.S. family renting their home, based on 2006 census data.

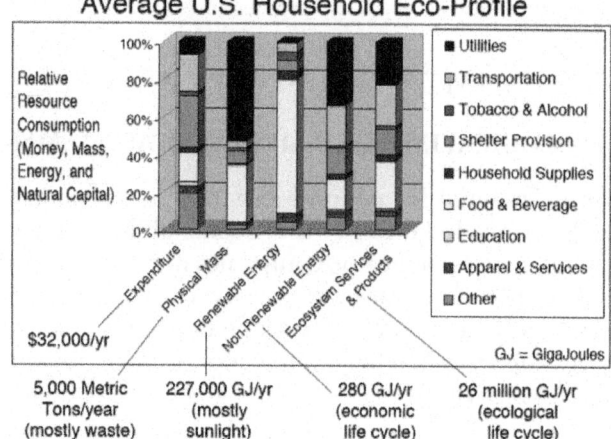

How Companies Can Use Eco-LCA™ to Support Business Decisions

Life cycle thinking is relevant to almost all business processes, from strategic planning to facility maintenance. The streamlined computational structure of Eco-LCA™ enables users to apply it repeatedly and test a variety of different business scenarios and assumptions. Thus, it can be used as a rapid screening tool to determine the merits of conducting a detailed LCA. The following are examples of how OSU has applied life cycle assessment to provide information for business and policy decisions:

New Product Development

Design teams can benefit from feedback about the life cycle implications of proposed designs. For example, OSU compared vapor-grown Carbon Nano-Fibers

(CNFs) to traditional materials such as aluminum, steel and polypropylene for auto body panels. The results indicate significantly higher life cycle energy use and environmental impacts of CNFs.

Energy Analysis

Concerns over energy costs and greenhouse gas (GHG) emissions have heightened the importance of LCA focused on energy use and GHG "footprints". For example, OSU conducted a landmark study comparing biofuels such as ethanol and biodiesel to fossil-based fuels. The results indicate the best "return on energy invested" for cellulosic ethanol derived from poplar and corn stover. While biofuels will reduce GHGs within the U.S., most other emissions will increase, and biofuels are more resource-intensive in terms of water, land, sand, stone, minerals, and even coal.

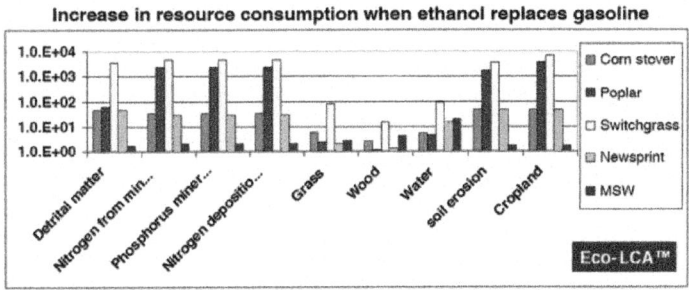

Supplier Management

Increased outsourcing and globalization raise the importance of monitoring supplier practices, including environmental performance. Eco-LCA™ can be used as a screening tool for "pareto" analysis of purchased products and services. For example, OSU has identified the top ten supplier sectors contributing to the life cycle GHG emissions footprint for soft drink manufacturing.

Critical Resource Planning

The continuity of global operations is dependent on the availability of key resources, including ecological products and services. To identify potential vulnerabilities, OSU has developed LCA-based indicators that can be applied to a company's global supply chain; for example, the "embedded water" index can highlight water-intensive operations in drought-prone regions.

Strategic Market Analysis

In an age of turbulence, markets will fluctuate due to changes in the cost and availability of natural resources. For an equipment manufacturer, OSU analyzed the dependency of its top customer sectors on threatened ecosystem services, including water, wood, nutrient cycling.

Product Benefit Claims

Increasingly, customers are seeking information about the environmental life cycle characteristics of products. Eco-LCA™ can be used in a rapid, iterative mode to explore potential product claims and to test competitor's claims. Then it can be applied in a detailed LCA to develop credible, scientific comparisons that account for the full spectrum of economic and ecological impacts.

Chapter 5

TRANSPORT MODEL FOR NANOFILTRATION AND REVERSE OSMOSIS SYSTEM BASED ON IRREVERSIBLE THERMODYNAMIC

REVERSE OSMOSIS WORKS

Reverse osmosis is one of the processes that makes desalination (or removing salt from seawater) possible. Beyond that, reverse osmosis is used for recycling, wastewater treatment, and can even produce energy.

Water issues have become an extremely pressing global threat. With climate change come unprecedented environmental impacts: torrential floodingin some areas, droughts in others, rising and falling sea levels. Add to that the threat of overpopulation -- and the demand and pollution a swelling population brings -- and water becomes one of the paramount environmental issues to watchfor in the next generation.

What is Reverse Osmosis?

Reverse Osmosis, a water treatment method traditionally known for removing salt from seawater, is also used to purify drinking water by forcing untreated water molecules through a semi-permeable membrane or filter. The membrane blocks contaminants and the impurities are subsequently expelled from the environment. The result is pure, clean drinking water.

Typical Reverse Osmosis System

1. Water containing impurities enters the system
2. Impurities are stopped and rejected at the membrane surface
3. Water pressure forces water molecules through the membrane

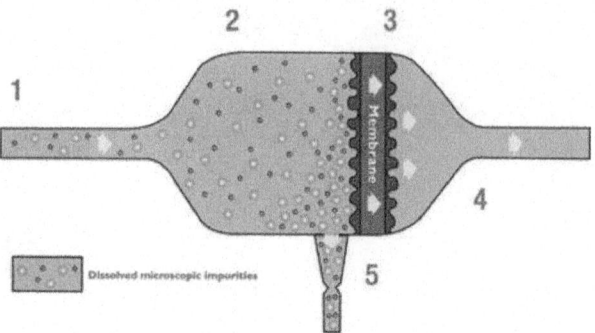

4. The purified water is then sent directly to the faucet

5. Impurities are expelled from the system and sent to a drain underneath the sink area.

The Reverse Osmosis membrane has a tight pore structure (less than 0.0001 micron or 500,000 times less than the diameter of a human hair) that effectively removes up to 99% of all contaminants and impurities such as total dissolved solids, chemicals, bacteria and viruses from drinking water. Anti-microbial filters used in Reverse Osmosis also help to remove unwanted odors, colors and tastes from water. Reverse Osmosis filtration technology is so effective that it is used by most leading water bottling plants.

What are some of the benefits?

- Pure, clean drinking water and ice cubes

- Removal of unwanted odors or tastes

- More robust tasting beverages that are mixed with Reverse Osmosis treated water (coffee, tea, *etc.*)

- Requires minimal maintenance

- Convenient- pure, clean water at the touch of a finger

- While not practical for purifying public water supply, an in-home Reverse Osmosis system can be cost effective when compared to other home water filtration methods

PureLux – A Unique Approach to Reverse Osmosis

A typical Reverse Osmosis filtration system consists of a sediment or chlorine removing pre-filter, a semi-permeable membrane, a storage tank and an activated carbon post-filter. However, these home water filtration systems only tend to operate at 5-20% efficiency, meaning for every ten gallons of water treated, only one gallon of pure water is achieved. Also, the holding tank used in many of the standard units can become a breeding ground for bacteria and other impurities after the water filtration process. The PureLux direct flow water purification system from Aerus is a unique Reverse Osmosis home water purification unit with five stages of filtration. There are three stages of carbon pre-filters to improve taste, remove sediment, organic and inorganic compounds. In stage four, water passes through the Reverse Osmosis membrane to remove impure substances and produce completely pure drinking water. In the fifth and final stage, water passes through an anti-microbial filter cartridge to prevent unpleasant odors, tastes and microorganisms from recurring as water leaves the faucet.

Benefits of PureLux Reverse Osmosis Filtration:

- Pure Water on Demand – No holding tank or storage unit, just fresh flowing water when you need it
- More Water, Less Waste – Runs at 75% efficiency, for every 10 gallons treated, 7.5 gallons of pure water are achieved, produces up to 300 gallons of purified water per day
- Smart System – Provides alerts for water quality, pressure leakage, filter capacity and replacement
- Energy Efficient – Consumes very little power
- Low Maintenance – Automated valves, pumps and cleaning, easy-to-read, user-friendly LCD panel, measures and reports Total Dissolved Solids (TDS)
- Eco-Friendly Design – Housing and filters are recyclable & biodegradable

Understanding Reverse Osmosis

Reverse Osmosis, commonly referred to as RO, is a process where you demineralize or deionize water by pushing it under pressure through a semi-permeable Reverse Osmosis Membrane.

Osmosis

To understand the purpose and process of Reverse Osmosis you must first understand the naturally occurring process of Osmosis.

Osmosis is a naturally occurring phenomenon and one of the most important processes in nature. It is a process where a weaker saline solution will tend to migrate to a strong saline solution. Examples of osmosis are when plant roots absorb water from the soil and our kidneys absorb water from our blood.

Below is a diagram which shows how osmosis works. A solution that is less concentrated will have a natural tendency to migrate to a solution with a higher concentration. For example, if you had a container full of water with a low salt concentration and another container full of water with a high salt concentration and they were separated by a semi-permeable membrane, then the water with the lower salt concentration would begin to migrate towards the water container with the higher salt concentration.

A semi-permeable membrane is a membrane that will allow some atoms or molecules to pass but not others. A simple example is a screen door. It allows air molecules to pass through but not pests or anything larger than the holes in the screen door. Another example is Gore-tex clothing fabric that contains an extremely thin plastic film into which billions of small pores have been cut. The pores are big enough to let water vapor through, but small enough to prevent liquid water from passing.

Osmosis

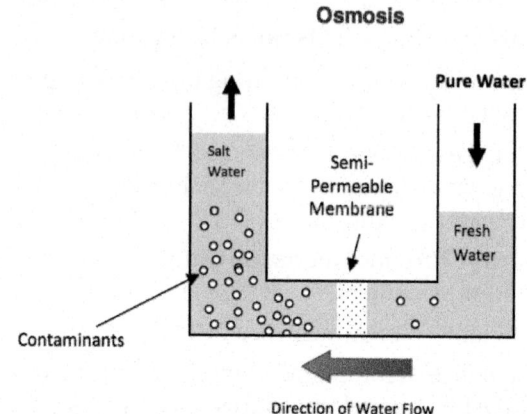

Direction of Water Flow

Reverse Osmosis is the process of Osmosis in reverse. Whereas Osmosis occurs naturally without energy required, to reverse the process of osmosis you need to apply energy to the more saline solution. A reverse osmosis membrane is a semi-permeable membrane that allows the passage of water molecules but not the majority of dissolved salts, organics, bacteria and pyrogens. However, you need to 'push' the water through the reverse osmosis membrane by applying pressure that is greater than the naturally occurring osmotic pressure in order to desalinate (demineralize or deionize) water in the process, allowing pure water through while holding back a majority of contaminants.

Below is a diagram outlining the process of Reverse Osmosis. When pressure is applied to the concentrated solution, the water molecules are forced through the semi-permeable membrane and the contaminants are not allowed through.

Reverse Osmosis

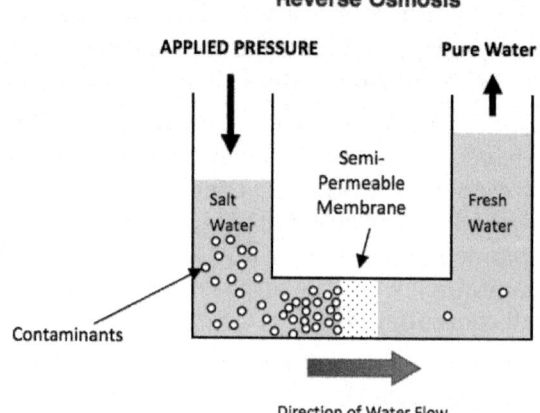

Direction of Water Flow

Function of Reverse Osmosis

Reverse Osmosis works by using a high pressure pump to increase the pressure on the salt side of the RO and force the water across the semi-permeable RO membrane, leaving almost all (around 95% to 99%) of dissolved salts behind in the reject stream. The amount of pressure required depends on the salt concentration of the feed water. The more concentrated the feed water, the more pressure is required to overcome the osmotic pressure.

The desalinated water that is demineralized or deionized, is called permeate (or product) water. The water stream that carries the concentrated contaminants that did not pass through the RO membrane is called the reject (or concentrate) stream.

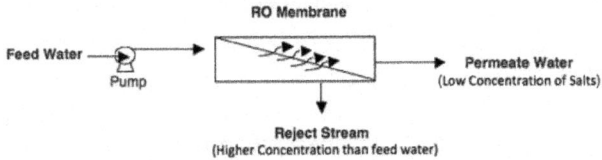

RO Membrane

Feed Water → Pump → | RO Membrane | → Permeate Water (Low Concentration of Salts)

Reject Stream (Higher Concentration than feed water)

As the feed water enters the RO membrane under pressure (enough pressure to overcome osmotic pressure) the water molecules pass through the semi-permeable membrane and the salts and other contaminants are not allowed to pass and are discharged through the reject stream (also known as the concentrate or brine stream), which goes to drain or can be fed back into the feed water supply in some circumstances to be recycled through the RO system to save water. The water that makes it through the RO membrane is called permeate or product water and usually has around 95% to 99% of the dissolved salts removed from it.

It is important to understand that an RO system employs cross filtration rather than standard filtration where the contaminants are collected within the filter media. With cross filtration, the solution passes through the filter, or crosses the filter, with two outlets: the filtered water goes one way and the contaminated water goes another way. To avoid build up of contaminants, cross flow filtration allows water to sweep away contaminant build up and also allow enough turbulence to keep the membrane surface clean.

IRREVERSIBLE PROCESS

In science, a process that is not reversible is called irreversible. This concept arises most frequently in thermodynamics.

In thermodynamics, a change in the thermodynamic state of a system and all of its surroundings cannot be precisely restored to its initial state by infinitesimal changes in some property of the system without expenditure of energy. A system that undergoes an irreversible process may still be capable of returning to its initial state; however, the impossibility occurs in restoring the environment to its own initial conditions. An irreversible process increases the entropy of the universe. However, because entropy is a state function, the change in entropy of a system is the same whether the process is reversible or irreversible. The second law of thermodynamics can be used to determine whether a process is reversible or not.

All complex natural processes are irreversible. The phenomenon of irreversibility results from the fact that if athermodynamic system, which is any system of sufficient complexity, of interacting molecules is brought from one thermodynamic state to another, the configuration or arrangement of the atoms and molecules in the system will change in a way that is not easily predictable. A certain amount of "transformation energy" will be used as the molecules of the "working body" do work on each other when they change from one state to another. During this transformation, there will be a certain amount of heat energy loss ordissipation due to intermolecular friction and collisions; energy that will not be recoverable if the process is reversed.

Many biological processes that were once thought to be reversible have been found to actually be a pairing of two irreversible processes. Whereas a single enzyme was once believed to catalyze both the forward and reverse chemical changes, research has found that two separate enzymes of similar structure are typically needed to perform what results in a pair of thermodynamically irreversible processes.

Absolute *Versus* Statistical Reversibility

Thermodynamics defines the statistical behaviour of large numbers of entities, whose exact behavior is given by more specific laws. Since the fundamental theoretical laws of physics are all time-reversible, however experimentally, probability of real reversibility is low, former presuppositions can be fulfilled and/ or former state recovered only to higher or lower degree. The irreversibility of thermodynamics must be statistical in nature; that is, that it must be merely highly unlikely, but not impossible, that a system will lower in entropy.

History

The German physicist Rudolf Clausius, in the 1850s, was the first to mathematically quantify the discovery of irreversibility in nature through his introduction of the concept ofentropy. In his 1854 memoir "On a Modified Form of the Second Fundamental Theorem in the Mechanical Theory of Heat" Clausius states:

> "It may, moreover, happen that instead of a descending transmission of heat accompanying, in the one and the same process, the ascending transmission, another permanent change may occur which has the peculiarity of *not being reversible* without either becoming replaced by a new permanent change of a similar kind, or producing a descending transmission of heat."

Simply, Clausius states that it is impossible for a system to transfer heat from a cooler body to a hotter body. For example, a cup of hot coffee placed in an area of room temperature (~72 °F) will transfer heat to its surroundings and thereby cool down with the temperature of the room slightly increasing (~72.3 °F). However, that same initial cup of coffee will never absorb heat from its surroundings causing it to grow even hotter with the temperature of the room decreasing (~71.7 °F). Therefore, the process of the coffee cooling down is irreversible unless extra energy is added to the system.

However, a paradox arose when attempting to reconcile microanalysis of a system with observations of its macrostate. Many processes are mathematically reversible in their microstate when analyzed using classical Newtonian mechanics. From 1872 to 1875, Ludwig Boltzmann reinforced the statistical explanation of this paradox in the form ofBoltzmann's entropy formula stating that as the number of possible microstates a system might be in increases, the entropy of the system increases and it becomes less likely that the system will return to an earlier state. His formulas quantified the work done by William Thomson, 1st Baron Kelvin who had argued that:

"The equations of motion in abstract dynamics are perfectly reversible; any solution of these equations remains valid when the time variable t is replaced by –t. Physical processes, on the other hand, are irreversible: for example, the friction of solids, conduction of heat, and diffusion. Nevertheless, the principle of dissipation of energy is compatible with a molecular theory in which each particle is subject to the laws of abstract dynamics."

Another explanation of irreversible systems was presented by French mathematician Henri Poincaré. In 1890, he published his first explanation of nonlinear dynamics, also called chaos theory. Applying the chaos theory to the second law of thermodynamics, the paradox of irreversibility can be explained in the errors associated with scaling from microstates to macrostates and the degrees of freedom used when making experimental observations. Sensitivity to initial conditions relating to the system and its environment at the microstate compounds into an exhibition of irreversible characteristics within the observable, physical realm.

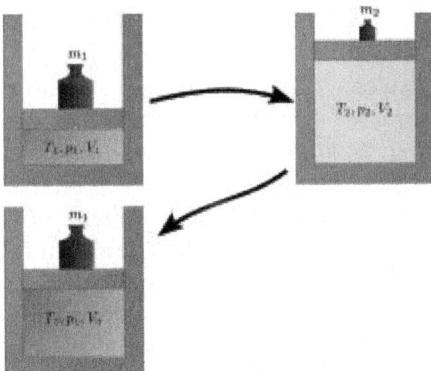

Fig. : Irreversible adiabatic process: If the cylinder is a perfect insulator, the initial top-left state cannot be reached anymore after it is changed to the one on the top-right. Instead, the state on the bottom left is assumed when going back to the original pressure because energy is converted into heat.

Examples of Irreversible Processes

In the physical realm, many irreversible processes are present to which the inability to achieve 100% efficiency in energy transfer can be attributed. The following is a list of spontaneous events which contribute to the irreversibility of processes.

- Heat transfer through a finite temperature difference
- Friction
- Plastic deformation
- Flow of electric current through a resistance
- Magnetization or polarization with a hysteresis
- Unrestrained expansion of fluids
- Spontaneous chemical reactions
- Spontaneous mixing of matter of varying composition/states

A Joule expansion is an example of classical thermodynamics, as it is easy to work out the resulting increase in entropy. It occurs where a volume of gas is kept in one side of a thermally isolated container (via a small partition), with the other side of the container being evacuated; the partition between the two parts of the container is then opened, and the gas fills the whole container. The internal energy of the gas remains the same, while the volume increases. The original state cannot be recovered by simply compressing the gas to its original volume, since the internal energy will be increased by this compression. The original state can only be recovered by then cooling the re-compressed system, and thereby irreversibly heating the environment. The diagram to the right applies only if the first expansion is "free" (Joule expansion). *i.e.* there can be no atmospheric pressure outside the cylinder and no weight lifted.

Complex Systems

The difference between reversible and irreversible events has particular explanatory value in complex systems (such as living organisms, or ecosystems). According to the biologists Humberto Maturana and Francisco Varela, living organisms are characterized by autopoiesis, which enables their continued existence. More primitive forms of self-organizing systems have been described by the physicist and chemist Ilya Prigogine. In the context of complex systems, events which lead to the end of certain self-organising processes, like death, extinction of a species or the collapse of a meteorological system can be considered as irreversible. Even if a clone with the same organizational principle (*e.g.* identical DNA-structure) could be developed, this would not mean that the former distinct system comes back into being. Events to which the self-organizing capacities of organisms, species or other complex systems can adapt, like minor injuries or changes in the physical environment are reversible. However, adaptation depends on import ofnegentropy into the organism, thereby increasing irreversible processes in its environment. Ecological principles, like those of sustainability and the precautionary principle can be defined with reference to the concept of reversibility.

IRREVERSIBLE THERMODYNAMICS

Irreversible thermodynamics is a division of physics which studies the general regularities in transport phenomena (heattransfer, mass transfer, *etc.*) and their relaxation (transition from nonequilibrium systems to the thermodynamical equilibrium state). It is possible to use for this purpose, as in reversible thermodynamics (thermostatics), phenomenological approaches based on the generalization of experimental facts and statistical physics methods which establish the bonds between molecular models (microscopical structure, properties of molecules, intermolecular interaction) and macroscopical substance behavior.

The starting points of irreversible thermodynamics are the *first* and *second laws of thermodynamics* in local formulation. In the general case, the nonequilibrium continuum is a moving v-component mixture of substances A_1, A_2,... A_v, among which r chemical reactions run simultaneously

$$\sum_{i=0}^{v} v'_{ik} A_i \rightleftharpoons \sum_{i=0}^{v} v''_{ik} A_i; 1 \le k \le r.$$

It is described by the fields of temperature $T(t, r)$, mass concentrations of components $\rho_i(t, r)$ and average mass flow speed $v(t, r)$. The local density of system $\rho(t, r)$ and mass fractions of components $x_i(t, r)$ are defined by the relations

$$\rho = \sum_{i=0}^{v} \rho'_{ik} X_i = \rho_i / \rho; \sum_{i=0}^{v} X_i = 1.$$

The conservation equation of mixture mass and balance equations of component masses in the local coordinate system, moving with the mass center speed $v(t, r)$, can be written as:

$$\frac{d\rho}{dt} = -\rho \frac{\partial}{\partial r} v; \rho \frac{dX_i}{dt} = -\frac{\partial}{\partial r} j_i + \sum_{k=1}^{r} m_i (v'_{ik} - v'_{ik}) R_k.$$

Here, j_i is the diffusion flow density of the ith component,

$$\sum_{i=1}^{r} j_i = 0,$$

and R_k is the rate of the kth chemical reaction. The balance equation of specific internal energy $u(t, r)$ has the following form:

$$\rho \frac{du}{dt} = -\frac{\partial}{\partial r} q - p \frac{\partial}{\partial r} v - \Pi : \frac{\partial v}{\partial r} + \sum_{i=1}^{v} F_i j_i,$$

where q is the heat flow density, p is pressure and Π is the viscosity part of pressure tensor

$$p = p1 + \Pi,$$

$\partial r / \partial r$ is the speed deformation tensor, F_i is the mass force exerted on the ith component particles from the external fields. Equation is the local formulation of the first law of thermodynamics for a non-equlibrium open system.

The second law of thermodynamics in the form of a local entropy balance is deduced by substituting the given equations with the thermodynamical with the given relationship

$$ds = \frac{1}{T}\left(du + pdv - \sum_{i=1}^{v}\mu_i d\xi_i\right),$$

describing the variation of local specific entropy s(t, r) with time interval dt in the center mass coordinate system of the continuum. In the given equation $v = 1/\rho$ is the specific volume, and $\mu_i(t, r)$ is the locai specific chemical potential of ith component. In so far as du in the given equation contains the dissipative contributions (the two last terms on the right side of the given equation); ds includes both the "equilibrium" component ds_e and the nonnegative term $ds_i \geq 0$; and conditioned by irreversibility:

$$ds = ds_e + ds_i; ds_i \geq 0.$$

The local entropy balance equation follows from the given equation:

$$\rho\frac{ds}{dt} = -\frac{\partial}{\partial r}j_s + \sigma,$$

where

$$j_s = \frac{1}{T}\left(q - \sum_{i=1}^{v}\mu_i j_i\right)$$

is entropy flow density, and

$$\sigma = -\sum_{i=1}^{v}j_i\left[\frac{\partial(\mu_i/T)}{\partial r} - \frac{1}{T}F_i\right]$$
$$-q\frac{1}{T^2}\frac{\partial T}{\partial r} - \Pi:\frac{1}{T}\frac{\partial v}{\partial r} - \sum_{i=1}^{v}R_k\frac{1}{T}A_k$$

is entropy production. In the given equation,

$$A_k = \sum_{i=1}^{v}m_i(v'_{ik} - v'_{ik})\mu_i$$

is the affinity of chemical reaction. The given equation Equation has the structure

$$\sigma = -\sum_{\alpha=1}^{n}J_\alpha X_\alpha,$$

where J_a are the flows (J_j, g, Π, R_k) and X_a are the matching thermodynamic forces. The second law of thermodynamics is reduced to

$$\sigma \geq 0.$$

Equality in the given equation corresponds to the thermodynamical equilibrium state — full or local, and nonequal — for the local entropy growth in irreversible processes. From the second law of thermodynamics) can be inferred the fact that flows in the given equation are in opposite direction to thermodynamic forces,

$$J_a X_a \leq 0$$

and they disappear after the transition to thermodynamic equilibrium. Practice fully confirms these deductions.

In accordance with experience, flows and thermodynamic forces are bound by transfer laws

$$J_a = F(X_j,..., X_n); 1 \leq a \leq n;$$

moreover if thermodynamic forces are absent, the flows are also equal to zero. The most simple relationship is linear,

$$J_\alpha = -\sum_{\beta=1}^{n} L_{\alpha\beta} X_\beta; 1 \leq \alpha \leq n.$$

The assumption about linearity of transport laws is a basis of linear, irreversible thermodynamics. Quantities $L_{\alpha\beta}$ are called kinetic (transport) coefficients. They are properties of the system considered; it means they don't depend on thermodynamic forces and flows, but are functions only of the state parameters (temperature, pressure, mixture composition). Phenomenological irreversible thermodynamics neither establishes these dependencies nor indicates suitable limits to linear transport laws. However, it is possible to formulate a number of general statements about the structure and properties of the kinetic coefficient matrix $L_{\alpha\beta}$.

From the given equations, it is clear that flows and thermodynamic forces have different tensor dimensions. Diffusion flows j_i and diffusion thermodynamic forces

$$d_i = \frac{\partial(\mu_i / T)}{\partial r} - \frac{1}{T} F_i$$

are vectors, as well as heat flow g and the related thermodynamic force $(1/T)\partial T/\partial r$. Reaction rates R_k and affinities A_k are scalars. *Some further scalar couples (J,X) are obtained from the last but one term in the given equation the normal and tangential viscosity tension contributions* are given by

$$\pi = (\Pi_{xx} + \Pi_{yy} + \Pi_{zz})/3; \overset{o}{\Pi} = \Pi - \pi_1,$$

and it also follows that

$$\Pi : \frac{\partial v}{\partial r} = \pi \frac{\partial}{\partial r} \cdot v + \overset{o}{\Pi} S; S = \frac{1}{2}\left(\frac{\partial v}{\partial r} + \frac{\partial \tilde{v}}{\partial r} \right) - \frac{1}{3} \frac{\partial}{\partial r} \cdot v$$

(S is the shift rate tensor, $\partial/\partial r$; $\equiv$ div). Thus π and $(1/T)(\partial/\partial r)\cdot v$ are the scalar flow and the thermodynamic force, respectively; flow $\overset{\circ}{\Pi}$ and force S/T are second-range tensors. According to the Courier principle, only flows and thermodynamic forces with the same tensor dimensions may be related using linear correlations. Any kinetic coefficient $L_{\alpha\beta}$, relating a flow J_α and a thermodynamic force X_β with different tensor dimensions is identically equal to zero. The permissible linear correlations between flows and thermodynamic forces occurring in non-equilibrium, multi-component reacting moving fluid are terms limited to

$$j_i = -\sum_{j=1}^{v} L_{ij} d_j - L_{iq} \frac{1}{T^2}\frac{\partial T}{\partial r}; 1 \le i \le v;$$

$$q = -\sum_{i=1}^{v} L_{qi} d_i - L_{qq} \frac{1}{T^2}\frac{\partial T}{\partial r};$$

$$\overset{\circ}{\Pi} = -L\frac{1}{T}S;$$

$$\pi = -L_{pp}\frac{1}{T}\frac{\partial}{\partial r}\cdot v - \sum_{k=1}^{r} L_{pk}\frac{1}{T}A_k;$$

$$R_k = -L_{bp}\frac{1}{T}\frac{\partial}{\partial r}\cdot v - \sum_{k=1}^{r} l_{kl}\frac{1}{T}A_1; 1 \le k \le r.$$

The correlations and the given equation predict not only direct ($J_\alpha = L_\alpha X_\alpha$) but "overcrossed" irreversible processes when the flow of a certain physical characteristic is implemented by other natural, "non-related" thermodynamic forces

$$(J_\alpha = L_{\alpha\beta}X_\beta; \alpha \ne \beta).$$

Thus the given equation is the *generalized Ficks law*, describes diffusion in a multi-component mixture when diffusion flow j_i of component i is caused by the diffusion thermodynamic forces d_i (in particular, by concentration gradients $\partial pj/\partial r$) of other components as well as temperature gradient. The quantities L_{ij} and L_{iq} represent the multi-component diffusion and thermal diffusion coefficients.

The generalized Fourier law describes a rise of heat flow at the expense of temperature gradient $\partial T/\partial r$ (the transport coefficient of this direct process is related to the ordinary thermal conductivity $\lambda = L_{qq}/T^2$), and by diffusion thermodynamic forces d_i. The latter process is called the diffusion thermo effect. It is in mutual overcross with thermal diffusion. The appearance of tangential viscous tensions, which is described by the generalized Newton-Stokes law, is a direct process (dynamic viscosity coefficient

$$\eta = [1/2][L/T])$$

and the overcross processes are absent here. Yet normal viscous tensions (at is clear from the given equation) can arise both because of the direct process — which stipulates volume expansion of the flow and is formed by the volume viscosity coefficient $\kappa = L_{pp}/T$ — and the overcross processes arising from nonequilibrium chemical reactions.

All the above irreversible processes have been observed in practice and are really linear until the fluid in which they proceed can be considered as a continuum.

It is impossible to say the same about the conditions of the given equation. The linearity of conditions between chemical reaction rates and their affinities is broken already by small deflections from chemical equilibrium $A_k = 0$, $R_k = 0$.

Uniting the given equation, it is possible to present entropy production in quadratic form

$$\sigma = \sum_{\alpha=1}^{n} \sum_{\beta=1}^{n} L_{\alpha\beta} X_\alpha X_\beta.$$

According to the given equation, it is positively determined. Hence it follows that all the diagonal elements of the kinetic coefficients matrix forming direct irreversible processes are positive, $L_{\alpha\alpha} \geq 0$; all its principal minors are positive too. Moreover, important information on the structure of this matrix is derived from Onzager mutuality correlations, which establish equality of kinetic coefficients for over-crossed, irreversible processes:

$$L_{\alpha\beta} = L_{\beta\alpha}.$$

The derivation of accurate formulas for kinetic coefficient calculation from molecular data, as well as molecular kinetic justification of the second law of thermodynamics for nonequilibrium systems in the form of the given equation, is feasible now only for dilute gases by applying the *Boltzman kinetic equation* and its modifications. Attempts to describe dense nonequilibrium system behavior proceeding from reversible, equations of molecular motion meets with problems that cannot be completely overcome. Together with the kinetic equations method, the nonequilibrium statistical operator method and related approaches of linear reaction theory can be successfully employed.

In certain cases, the introduction of some critical nonequilibrity degree can suddenly increase an open system's regularity, which is followed by dissipative structure formation. Known examples are: the occurrence of Benard cells in a viscous liquid-layer, heated from below in gravitational field, when the heat flux exceeds a critical value; the appearance of layers in a gas discharge column with the introduction of a critical current density; and so on. A richer variety of dissipative structures — both space and temporary — are found in nonlinear, noneqilibrium systems (for example, the Belousov — Zabotinski reaction). A comparatively new and swiftly-developing branch of irreversible thermodynamics is devoted to the study of the organization of phenomena in strong nonequilibrium, nonlinear open systems.

REVERSIBLE AND IRREVERSIBLE THERMODYNAMIC PROCESSES

We have seen above that in absence of any gradients (or motive forces) a thermodynamic system continues to remain in a state of equilibrium. Obviously, if a disturbance (*i.e.*, mechanical, thermal or chemical potential gradient) is impressed upon such a system it will transit from its initial state of equilibrium. However, as it moves away from its initial state the originally applied gradients will diminish progressively in time, and ultimately when they are reduced to infinitesimal levels the system will attain a new equilibrium state. A question arises here as to the nature of the process of change: if the initially impressed disturbances are reversed in direction (not magnitude) can the system return to its first equilibrium state back through the same intermediate states as it went through during the first phase of change? If that happens we depict the process as *reversible*, if not, then the process is termed *irreversible*.

It is necessary to understand the concept of *reversibility* of thermodynamic process more deeply as it is an *idealized* form of process of change and without that consideration it is not possible to represent or understand real thermodynamic processes, which are generally irreversible in nature.

What makes a thermodynamic process reversible? To answer the question let us again take the example of the simple gas-in-piston-and-cylinder system.

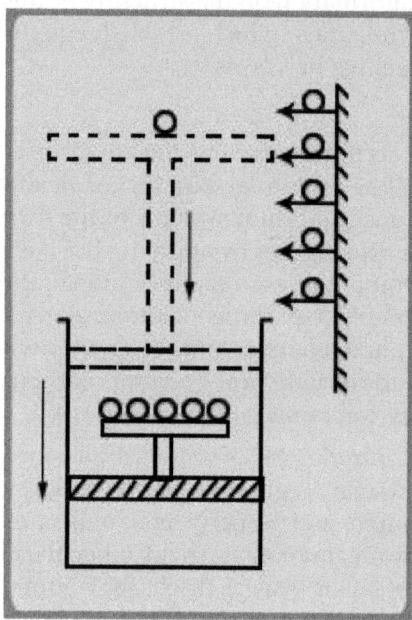

Fig. : Illustration of Reversibility of Thermodynamic Process.

The system initially contains a pure gas whose pressure equals that exerted externally (due to piston weight), and its temperature is the same as that of the environment. Thus it is at equilibrium (say state 'A') as there are no mechanical, thermal or chemical concentration gradients in the system. Now a ball of a known weight is transferred on to the piston, whereupon the external pressure exceeds the gas pressure and the piston moves down to attain a new lower position at which point the gas has been compressed and its pressure once again equals that applied externally. At the same time if any differentials in temperature (within or across the system boundary) and internal concentration distribution of the gas molecules result due to the applied mechanical imbalance, heat and mass transfer will take place simultaneously until these gradients are also annulled and the system eventually comes to rest at a new equilibrium point (say, 'B'). We say that the system has undergone a process due to which its state has changed from A to B. Note that this process can be continued as long as desires by sequentially transferring more and more balls individually onto the piston and impelling the system to change in steps till say the end point state 'X'. The question that one may pose: is the process A-X reversible? That is, if one reversed all the initial steps of sequentially moving each ball off the piston so as to reach from state 'X' back to 'A' would all the interim states of the system as defined by temperature, pressure and volume at any point be identical to those obtained during the process of going from A to X? To answer this question we need to understand the process occurring in the system a little more deeply. Consider first that a mass mo is suddenly moved onto the piston from a shelf (at the same level). The piston assembly accelerates downwards, reaching its maximum velocity at the point where the downward force on the piston just balanced by the pressure exerted by the gas in the cylinder. However, the initial momentum of the plunging piston would carry it to a somewhat lower level, at which point it reverses direction. If the piston were held in this position of maximum depression brought about by transfer of the mass m_0, the decrease in its potential-energy would very nearly equal the work done on the gas during the downward movement. However, if unrestrained, the piston assembly would oscillate, with progressively decreasing amplitude, and would eventually come to rest at a new equilibrium position at a level below its initial position.

The oscillation of the piston assembly cease because it is opposed by the viscosity of the gas, leading to a gradual conversion of the work initially done by the piston into heat, which in turn is converted to internal energy of the gas.

All processes carried out in finite time with real substances are accompanied in some degree by dissipative effects of one kind or another. However, one may conceive of processes that are free of dissipative effects. For the compression process, such effects issue from sudden addition of a finite mass to the piston. The resulting imbalance of forces acting on the piston causes its acceleration, and leads to its subsequent oscillation. The sudden addition of smaller mass increments may reduce but does not eliminate this dissipative effect. Even the addition of an infinitesimal mass leads to piston oscillations of infinitesimal amplitude and a consequent dissipative effect. However, one may *conceive* of an ideal process in which small mass increments are added one after another at a rate such that the piston movement downwards is continuous, with minute oscillation only at the end of the entire process.

This *idealized* case derives if one imagines of the masses added to the piston as being infinitesimally small. In such a situation the piston moves down at a uniform but infinitesimally slow rate. Since the disturbance each time is infinitesimal, the system is always infinitesimally displaced from the equilibrium state both internally as well with respect to external surroundings. Such a process which occurs very slowly and with infinitesimal driving forces is called a *quasi-static* process. To freeze ideas let us assume that the gas in the system follows the ideal gas law. Thus the pressure, temperature and volume at any point during the process are related by eqn. 1.12 (or 1.13). Now imagine that the process of gradual compression is reversed by removing each infinitesimal mass from the piston just as they were added during the forward process. Since during the expansion process also the system will always be differentially removed from equilibrium state at each point, the pressure, temperature and volume will also be governed by the relation 1.12. Since the latter is an equilibrium relationship and hence a unique one, each interim state of the system would exactly converge during both forward and backward progress of system states. Under such a condition the process of compression is said to be thermodynamically reversible. Both the system and its surroundings are ultimately restored to their initial conditions. In summary, therefore, if both the system and its surroundings can be restored to their respective initial states by reversing the direction of the process, then the process is said to be reversible. If a process does not fulfill this criterion it is called an *irreversible* process.

It need be emphasized that a reversible process need be a quasi-static process, and that the origin of irreversibility lie in the existence of dissipative forces in real systems, such as viscosity, mechanical friction. These forces degrade *useful* work irreversibly into heat which is not re-convertible by simply reversing the direction of the process, since during a reverse process a fraction of useful work will again be lost in the form heat in overcoming the dissipative forces. Thus, in the above example system if there were no viscous or frictional forces opposing the motion of the piston the processes of compression and expansion would be reversible, provided of course all changes occur under infinitesimal gradients of force. The argument in the last sentence may be extended to state that if changes are brought about by finite gradients (in this case finite difference in force across the piston, associated with addition of finite mass to the piston), the process would necessarily be irreversible. This is because finite gradients will force the system to traverse through non-equlibrium interim states, during which the pressure, temperature and volume will not be constrained by a unique relationship such as eqn. 1.12, which holds for equilibrium states. Indeed, it would not be possible to define the non-equilibrium states in terms of a single temperature or pressure, as there would be internal gradients of these variables during processes induced by finite force imbalances across the system boundary. These very same considerations would apply for the reverse process of expansion as well, if it occurs under finite mechanical gradients. So in general during such processes it would not be possible to ascribe unique intensive properties to interim states during a change, and hence the forward and reverse "paths" would not coincide as they would if the process occurs under quasi-static conditions.

An additional point that obtains from the above considerations is that only under reversible conditions can one calculate the thermodynamic work by integrating equation, since at all points during the process the variables P, V and T are always uniquely related by the equation. Clearly if the process were occurring under irreversible conditions no such relation would hold and hence the calculation of the thermodynamic work would not be possible through a simple integration of equation.

The foregoing discussion has used the example of a single-phase closed-system, where compression and expansion processes are induced by gradients of mechanical force across the system boundary. There are, however, many processes which are occur due to potential gradients other than mechanical forces. For example, heat flow is induced by temperature differences, electromotive force gradients lead to flow of electricity, and chemical reactions take place as there is a difference between the chemical potential of reactants and products. In general, it may be shown that all such processes brought about by potential gradients of various kinds would tend to reversibility if the gradients are themselves infinitesimal. For example, heat transfer across the boundary of a thermodynamic system would be reversible if the difference across it is of a differential amount $'dT'$, and so on.

FUNDAMENTAL OF IRREVERSIBLE THERMODYNAMICS IN MEMBRANE SYSTEM

Membrane Transport

In cellular biology, membrane transport refers to the collection of mechanisms that regulate the passage of solutes such as ions and small molecules through biological membranes, which are lipid bilayers that contain proteins embedded in them. The regulation of passage through the membrane is due to selective membrane permeability - a characteristic of biological membranes which allows them to separate substances of distinct chemical nature. In other words, they can be permeable to certain substances but not to others.

The movements of most solutes through the membrane are mediated by membrane transport proteins which are specialized to varying degrees in the transport of specific molecules. As the diversity and physiology of the distinct cells is highly related to their capacities to attract different external elements, it is postulated that there is a group of specific transport proteins for each cell type and for every specific physiological stage. This differential expression is regulated through the differential transcription of the genescoding for these proteins and its translation, for instance, through genetic-molecular mechanisms, but also at the cell biology level: the production of these proteins can be activated by cellular signaling pathways, at the biochemical level, or even by being situated in cytoplasmic vesicles.

Background

Thermodynamically the flow of substances from one compartment to another can occur in the direction of a concentration or electrochemical gradient or against it. If the exchange of substances occurs in the direction of the gradient, that is, in the direction of decreasing potential, there is no requirement for an input of energy from outside the system; if, however, the transport is against the gradient, it will require the input of energy, metabolic energy in this case. For example, a classic chemical mechanism for separation that does not require the addition of external energy is dialysis. In this system a semipermeable membrane separates two solutions of different concentration of the same solute. If the membrane allows the passage of water but not the solute the water will move into the compartment with the greatest solute concentration in order to establish an equilibrium in which the energy of the system is at a minimum. This takes place because the water moves from a high solvent concentration to a low one (in terms of the solute, the opposite occurs) and because the water is moving along a gradient there is no need for an external input of energy.

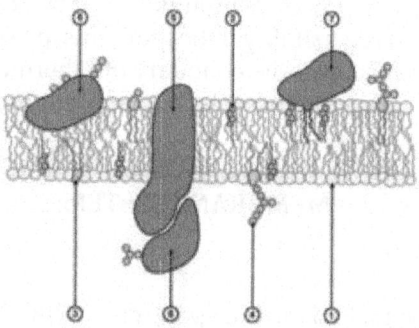

Fig. : Diagram of a cell membrane.

1. phospholipid 2. cholesterol 3. glycolipid 4. sugar 5. polytopic protein (transmembrane protein) 6. monotopic protein (here, a glycoprotein) 7. monotopic protein anchored by a phospholipid 8. peripheral monotopic protein (here, a glycoprotein.)

The nature of biological membranes, especially that of its lipids, is amphiphilic, as they form bilayers that contain an internalhydrophobic layer and an external hydrophilic layer. This structure makes transport possible by simple or passive diffusion, which consists of the diffusion of substances through the membrane without expending metabolic energy and without the aid of transport proteins. If the transported substance has a net electrical charge, it will move not only in response to a concentration gradient, but also to an electrochemical gradient due to the membrane potential.

Table : Relative permeability of a phospholipid bilayer to various substances.

Type of substance	Examples	Behaviour
Gases	CO_2, N_2, O_2	Permeable
Small uncharged polar molecules	Urea, water, ethanol	Permeable, totally or partially
Large uncharged polar molecules	glucose, fructose	Not permeable
Ions	K^+, Na^+, Cl^-, HCO_3^-	Not permeable
Charged polar molecules	ATP, amino acids, glucose-6-phosphate	Not permeable

As few molecules are able to diffuse through a lipid membrane the majority of the transport processes involve transport proteins. These transmembrane proteins possess a large number of alpha helices immersed in the lipid matrix. In bacteria these proteins are present in the beta lamina form. This structure probably involves a conduit through hydrophilic protein environments that cause a disruption in the highly hydrophobic medium formed by the lipids. These proteins can be involved in transport in a number of ways: they act as pumps driven by ATP, that is, by metabolic energy, or as channels of facilitated diffusion.

Thermodynamics

A physiological process can only take place if it complies with basic thermodynamic principles. Membrane transport obeys physical laws that define its capabilities and therefore its biological utility.

A general principle of thermodynamics that governs the transfer of substances through membranes and other surfaces is that the exchange of free energy, ΔG, for the transport of a mole of a substance of concentration C_1 in a compartment to another compartment where it is present at C_2 is:

$$\Delta G = RT \log \frac{C_2}{C_1}$$

Where C_2 is less than C_1 ΔG is negative, and the process is thermodynamically favorable. As the energy is transferred from one compartment to another, except where other factors intervene, an equilibrium will be reached where $C_2 = C_1$, and where $G = 0$. However, there are three circumstances under which this equilibrium will not be reached, circumstances which are vital for the *in vivo* functioning of biological membranes:

- The macromolecules on one side of the membrane can bond preferentially to a certain component of the membrane or chemically modify it. In this way, although the concentration of the solute may actually be different on both sides of the membrane, the availability of the solute is reduced in one of the compartments to such an extent that, for practical purposes, no gradient exists to drive transport.

- A membrane electrical potential can exist which can influence ion distribution. For example, for the transport of ions from the exterior to the interior, it is possible that:

$$\Delta G = RT \log \frac{C_{inside}}{C_{outside}} + ZF\Delta P$$

Where F is Faraday's constant and ΔP the membrane potential in volts. If ΔP is negative and Z is positive, the contribution of the term $ZF\Delta P$ to ΔG will be negative, that is, it will favor the transport of cations from the interior of the cell. So, if the potential difference is maintained, the equilibrium state $\Delta G = 0$ will not correspond to an equimolar concentration of ions on both sides of the membrane.

- If a process with a negative ΔG is coupled to the transport process then the global ΔG will be modified. This situation is common in active transport and is described thus:

$$\Delta G = RT \log \frac{C_{inside}}{C_{outside}} + \Delta G^{b}$$

Where ΔG^{b} corresponds to a favorable thermodynamic reaction, such as the hydrolysis of ATP, or the co-transport of a compound that is moved in the direction of its gradient.

Transport Types

Passive Diffusion

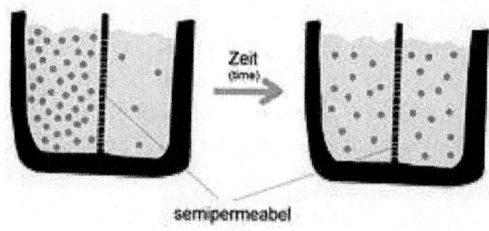

semipermeabel

Fig. : A semipermeable membrane separates two compartments of different solute concentrations: over time, the solute will diffuse until equilibrium is reached.

The passive diffusion is a spontaneous phenomenon that increases the entropy of a system and decreases the free energy. The transport process is influenced by the characteristics of the transport substance and the nature of the bilayer. Membrane proteins (with the exception of channels - facilitated diffusion) are not involved in passive diffusion. The diffusion velocity of a pure phospholipid membrane will depend on:

- concentration gradient,
- hydrophobicity,
- size,
- charge, if the molecule has a net charge.

Active and Co-transport

In active transport a solute is moved against a concentration or electrochemical gradient, in doing so the transport proteins involved consume metabolic energy, usually ATP. Inprimary active transport the hydrolysis of the energy provider (*e.g.* ATP) takes place directly in order to transport the solute in question, for instance, when the transport proteins are ATPase enzymes. Where the hydrolysis of the energy provider is indirect as is the case in secondary active transport, use is made of the energy stored in an electrochemical gradient. For example, in co-transport use is made of the gradients of certain solutes to transport a target compound against its gradient, causing the dissipation of the solute gradient. It may appear that, in this example, there is no energy use, but hydrolysis of the energy provider is required to establish the gradient of the solute transported along with the target compound. The gradient of the co-transported solute will be generated through the use of certain types of proteins called biochemical pumps.

The discovery of the existence of this type of transporter protein came from the study of the kinetics of cross-membrane molecule transport. For certain solutes it was noted that the transport velocity reached a plateau at a particular concentration above which there was no significant increase in uptake rate, indicating a log curve type response. This was interpreted as showing that transport was mediated by the formation of a substrate-transporter complex, which is conceptually the same as the enzyme-substrate complex ofenzyme kinetics. Therefore, each transport protein has an affinity constant for a solute that is equal to the concentration of the solute when the transport velocity is half its maximum value. This is equivalent in the case of an enzyme to the Michaelis-Menten constant.

Some important features of active transport in addition to its ability to intervene even against a gradient, its kinetics and the use of ATP, are its high selectivity and ease of selective pharmacological inhibition

Secondary Active Transporter Proteins

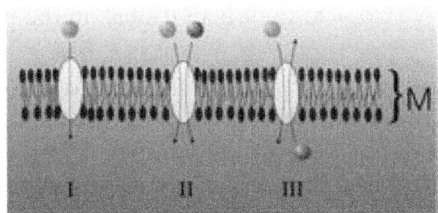

Fig. : Uniport, symport, and antiport of molecules through membranes.

Secondary active transporter proteins move two molecules at the same time: one against a gradient and the other with its gradient. They are distinguished according to the directionality of the two molecules:

- antiporter: (also called exchanger or counter-transporter) move a molecule against its gradient and at the same time displaces one or more ions along its gradient. The molecules move in opposite directions.
- symporter:move a molecule against its gradient while displacing one or more different ions along their gradient. The molecules move in the same direction.

Both can be referred to as co-transporters.

Pumps

A pump is a protein that hydrolyses ATP in order to transport a particular solute through a membrane in order to generate an electrochemical gradient to confer certain membrane potential characteristics on it. This gradient is of interest

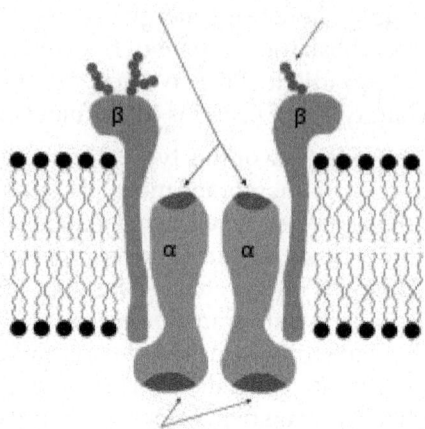

Fig. : Simplified diagram of a sodium potassium pump showing alpha and beta units.

as an indicator of the state of the cell through parameters such as the Nernst potential. In terms of membrane transport the gradient is of interest as it contributes to increased system entropy in the co-transport of substances against their gradient. One of the most important pumps in animal cells is thesodium potassium pump, that operates through the following mechanism:

1. Binding of three Na^+ ions to their active sites on the pump which are bound to ATP.

2. ATP is hydrolyzed leading to phosphorylation of the cytoplasmic side of the pump, this induces a structure change in the protein. The phosphorylation is caused by the transfer of the terminal group of ATP to a residue of aspartate in the transport protein and the subsequent release of ADP.

3. The structure change in the pump exposes the Na^+ to the exterior. The phosphorylated form of the pump has a low affinity for Na^+ions so they are released.

4. Once the Na^+ ions are liberated, the pump binds two molecules of K^+ to their respective bonding sites on the extracellular face of the transport protein. This causes the dephosphorylation of the pump, reverting it to its previous conformational state, transporting the K^+ ions into the cell.

5. The unphosphorylated form of the pump has a higher affinity for Na^+ ions than K^+ ions, so the two bound K^+ ions are released into the cytosol. ATP binds, and the process starts again.

Membrane Selectivity

As the main characteristic of transport through a biological membrane is its selectivity and its subsequent behavior as a barrier for certain substances, the underlying physiology of the phenomenon has been studied extensively. Investigation into membrane selectivity have classically been divided into those relating to electrolytes and non-electrolytes.

Electrolyte Selectivity

The ionic channels define an internal diameter that permits the passage of small ions that is related to various characteristics of the ions that could potentially be transported. As the size of the ion is related to its chemical species, it could be assumed *a priori* that a channel whose pore diameter was sufficient to allow the passage of one ion would also allow the transfer of others of smaller size, however, this does not occur in the majority of cases. There are two characteristics alongside size that are important in the determination of the selectivity of the membrane pores: the facility for dehydration and the interaction of the ion with the internal charges of the pore.

In order for an ion to pass through a pore it must dissociate itself from the water molecules that cover it in successive layers of solvation. The tendency to dehydrate, or the facility to do this, is related to the size of the ion: larger ions can do it more easily that the smaller ions, so that a pore with weak polar centres will preferentially allow passage of larger ions over the smaller ones. When the interior of the channel is composed of polar groups from the side chains of the component amino acids, the interaction of a dehydrated ion with these centres can be more important than the facility for dehydration in conferring the specificity of the channel. For example, a channel made up of histidines and arginines, with positively charged groups, will selectively repel ions of the same polarity, but will facilitate the passage of negatively charged ions. Also, in this case, the smallest ions will be able to interact more closely due to the spatial arrangement of the molecule (stericity), which greatly increases the charge-charge interactions and therefore exaggerates the effect.

Non-electrolyte Selectivity

Non-electrolytes, substances that generally are hydrophobic and lipophylic, usually pass through the membrane by dissolution in the lipid bilayer, and therefore, by passive diffusion. For those non-electrolytes whose transport through the membrane is mediated by a transport protein the ability to diffuse is, generally, dependent on the partition coefficient K. Partially charged non-electrolytes, that are more or less polar, such as ethanol, methanol or urea, are able to pass through the membrane through aqueous channels immersed in the membrane. It is interesting to note that there is no effective regulation mechanism that limits this transport, which indicates an intrinsic vulnerability of the cells to the penetration of these molecules.

Creation of Membrane Transport Proteins

There are several databases which attempt to construct phylogenetic trees detailing the creation of transporter proteins. One such resource is the Transporter Classification database.

PASSIVE TRANSPORT

Passive transport is the simplest method of transport and is dependent upon the concentration gradient, and the size and charge of the solute. In passive transport, small uncharged solute particles diffuse across the membrane until both sides of the membrane have reached an equilibrium that is similar in concentration. The direction of solute travel is indicative of the concentration of that particular particle on each side of the membrane.

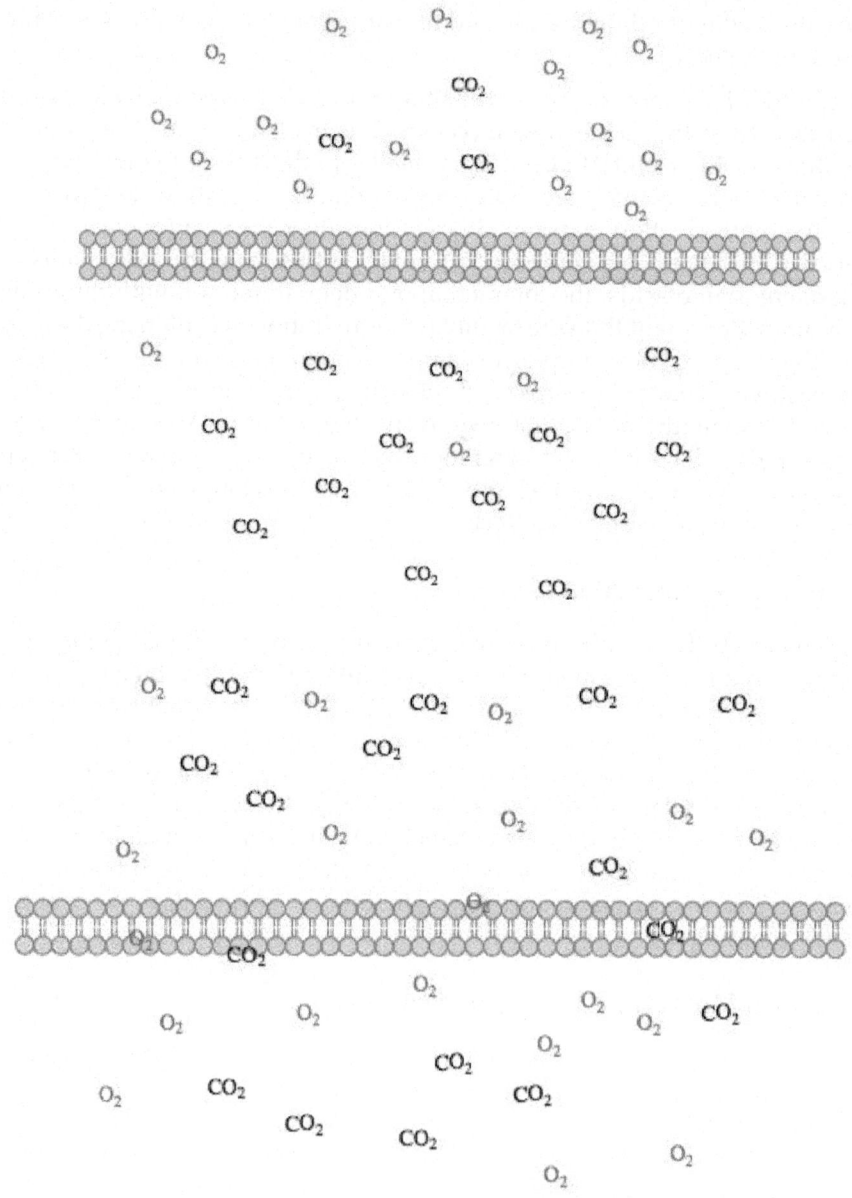

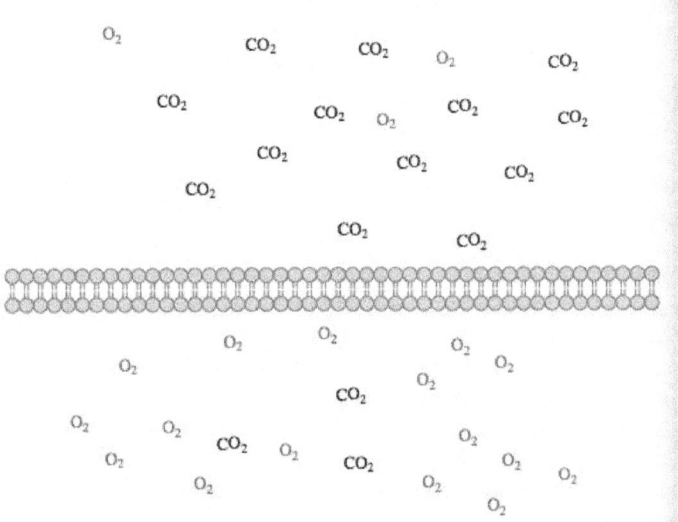

Fig. : Passive diffusion of O_2 and CO_2 across a membrane over time.

Based on the thermodynamics of the system, particles will move from an area of high concentration to an area of low concentration in order to increase the entropy of the cell. Additionally, this particle movement will occur spontaneously as the free energy (Gibbs free energy; ΔG) of the system is negative.

$$G = H - TS$$

Where:

- G = Gibbs free energy (Joules)
- H = Enthalpy (Joules)
- T = Absolute temperature (°K)
- S = Entropy (Joules/°K)

Further, the amount of energy consumed or released by the system is as follows:

$$\Delta G = RT \log \frac{C_2}{C_1}$$

Where:

- ΔG = Change in Gibbs free energy (Joules)
- R = Gas Constant (Joules/((Moles)(°K)))
- C_1 = Extracellular concentration of solute (Moles)
- C_2 = Intracellular concentration of solute (Moles)

If ΔG < 0, then particle movement is considered to be spontaneous; whereas, ΔG > 0 particle movement requires the input of energy to move in the desired direction.

The properties of the membrane must also be considered when determining the rate of flow of the substrate. Darcy's Law can be used to determine flow rate.

$$Q = \left(\frac{K}{\eta}\right)\left(\frac{\Delta P}{\Delta L}\right) A$$

Where:

- Q = flow rate (m³/s)
- K = permeability (m²)
- ΔP = pressure difference (Pa)
- ΔL = flow length or test sample thickness (m)
- A = cross-sectional area of flow (m²)
- η = fluid viscosity (Pa s)

The permeability coefficient is dependent upon the porous nature of the membrane and fluid.

$$P = \frac{KD}{\Delta X}$$

Where:

- P = permeability coefficient (cm/s)
- K = partition coefficient
- D = diffusion coefficient (cm²/s)
- ΔX = width of membrane (cm)

Common permeability coefficients are displayed below:

Permeability Coefficient (μm/s)	Membrane	Substrate
300	Escherichia coli	Glycerol
25	Escherichia coli	Glucose
1	Escherichia coli	Lactose

The diffusion of small charged particles, on the other hand, across a membrane is dependent upon the charge and transmembrane concentration of the solute. Again, however, the direction of solute travel is indicative of the thermodynamics of the system. Particles will travel from an area of high to low concentration, as well as travel so that the electrical potential across the membrane is diminished. As a result of this movement, the entropy of the system has increased.

Passive transport is independent of membrane proteins and the catabolism of biological molecules for energy. This energy deficient process commonly occurs in the blood brain barrier as specific molecules, such as sodium thiopental, can diffuse across the membrane. Sodium thiopental is a barbiturate frequently used in the methods of lethal injection. Sodium thiopental is a negatively charged particle and proceeds across the blood brain barrier to neuronal synaptic clefts. Sodium thiopental is an agonist of γ-aminobutyric acid (GABA), which acts as a neurotransmitter inhibitor. Sodium thiopental acts an anesthetic to cause unconsciousness.

Additionally, passive diffusion occurs across the placenta as all solute particles are exchanged between mother and fetus. Placental physiology juxtaposes maternal and fetal capillaries in order to exchange solute particles, such as oxygen and carbon dioxide gas. These uncharged molecules proceed across maternal and fetal capillary membranes in the direction from high to low concentration. This spontaneous process occurs in accord with the aforementioned properties. Despite the cellular system, passive diffusion across the membranes of all biological cells.

Facilitated Diffusion

Facilitated diffusion, not to be confused with simple diffusion, is a form of passive transport mediated by transport proteins imbedded within the cellular membrane. Facilitated diffusion allows the passage of lipophobic molecules through the cell membrane's lipid bilayer. Just as in passive transport, molecules, particles, and ions travel freely across the cellular membrane from high concentration to low concentration in an attempt to achieve equilibrium and thereby increase the entropy of the system. Also like passive transport, the Gibbs Free Energy of the system is negative, allowing the particle movement to be spontaneous. Facilitated diffusion, however, uses channel proteins to facilitate solute movement.

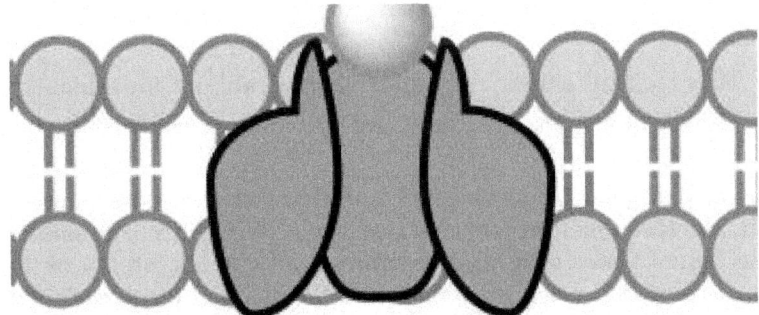

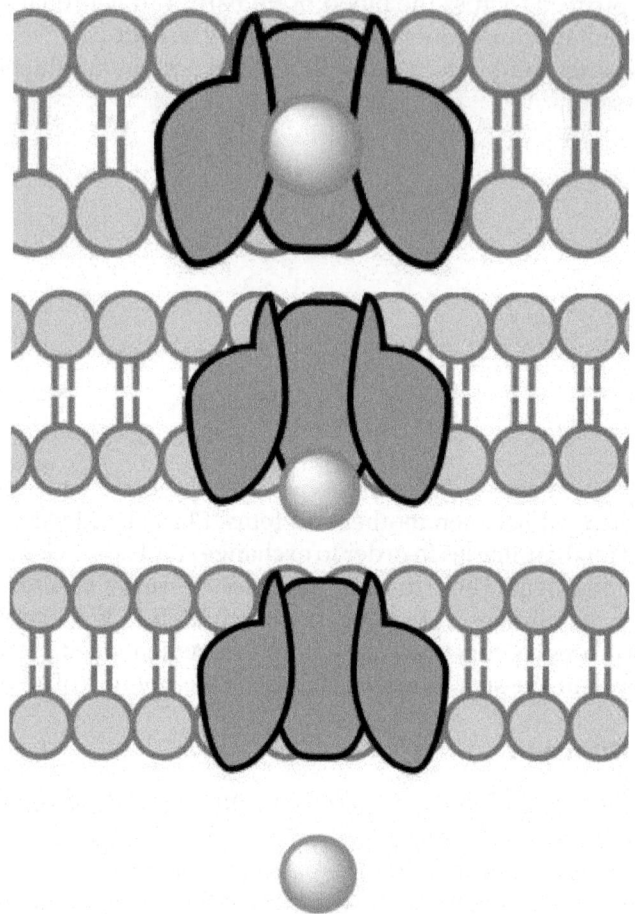

Fig. : Facilitated Diffusion via channel protein across a membrane proceeding images.

Channel Proteins

Channel proteins are pores immersed in the lipid bilayer membrane and are the hallmark of facilitated diffusion. All channel proteins have two things in common:

1. They facilitate a thermodynamically favorable net movement of particles
2. They demonstrate an affinity and specificity for the particle being transported.

It is of particular importance to note that channel proteins are not indiscriminate; each channel protein contains a selectivity filter. The selectivity filter is a collection of amino acid residues concentrated in the interior of the channel protein. As particles, often ions, pass into the channel protein, an electrostatic interaction occurs between the amino acid residues and the ion. The interaction would, for example, involve negatively charged amino acid residues in the case of ions like calcium (Na^+) or potassium (K^+), and positively charged amino acid residues in the case of chlorine (Cl^-). The electrostatic interaction between amino acid residues and ions allows the channel protein to identify the ion in question by measuring its atomic radius with extremely finite accuracy. Potassium (K^+) channels select K^+ over Na^+ by a factor of over one thousand despite differing in atomic radius by a mere $0.38\,\mathring{A}$.

While all channel proteins have an inherent selectivity filter, others have additional gating. Gating is a response to a predetermined trigger that allows the channel protein to undergo a conformational change. This action subsequently causes another conformational change that either opens or closes the channel, allowing or disallowing its specific particle to pass. Channel proteins can be physically or chemically modulated through a number of different mechanisms.

Voltage Gating

Voltage gated channel proteins are activated by a change in the electrical potential of the cellular membrane in its vicinity. When a potential difference occurs across the cellular membrane, its electromagnetic field causes a conformational change in the channel protein, allowing it to open. Opening of the protein channel allows an influx or efflux of ions which, in turn, depolarizes the cell membrane.

Voltage gated protein channels play a particularly important role in excitable neuronal and muscle tissues.

Ligand Gating

Ligand gated channel proteins are activated in response to the binding of a ligand. Typically, ligand binding occurs at an allosteric binding site independent of the channel protein's pore. The binding of a ligand at the allosteric binding site causes a conformational change in the structure of the channel protein, subsequently causing an influx or efflux of ions. Release of the ligand allows the channel protein to return to its original shape. Structurally, ligand gated channel proteins generally differ from other channels due to the presence of an additional protein domain that serves as the allosteric binding site.

The prototypical example of ligand gating is the nicotinic acetylcholine receptor located on the postsynaptic side of the neuromuscular junction.

Other Gating

Channel proteins may be gated in less common instances by methods such as light activation, mechanical activation, or secondary messanger activation. Light activated protein channels contain a photoswitch through which a photon causes a conformational change in the channel protein causing it to open or close. Only one such protein channel exists naturally. Mechanically activated protein channels open or close in response to a mechanical stimulus and are vital to the touch, hearing, and balance sensations in human. Ligand-gated protein channels are typically linked to second messanger gating. Second messenger gating functions stepwise in that a neurotransmitter binds to a channel protein receptor which, in turn, reveals an active site to which the conformation-changing ligand binds.

Active Transport

Active transport, simply put, is the movement of particles through a transport protein from low concentration to high concentration at the expense of metabolic energy. The most common energy source used by cells is adenosine triphosphate or ATP, though other sources such as light energy or the energy stored in an electrochemical gradient are also utilized. In the case of ATP, energy is chemically harvested through hydrolysis.ATP hydrolysis in turn causes a conformational change in the transport protein which allows mechanical movement of the particle in question. Active transport systems are, therefore, energy-coupling devices as chemical and mechanical processes are linked to achieve particle movement. Active transport is classified as either Primary Active Transport or Secondary Active Transport. A ribbon structure of a commonly depicted ABC Vitamin B_{12}importer active transport protein.

Primary Active Transport

Primary active transport uses the energy found in ATP, photons, and electrochemical gradients directly in the transport of molecules from low concentration to high concentration across the cellular membrane.

Using ATP

The enzyme-catalyzed hydrolysis reaction removing a phosphate from ATP, thereby forming ADP, causes a conformational change in the transport protein allowing particles to influx or efflux. Enzymes catalyzing ATP-driven primary active transport are called ATPases.

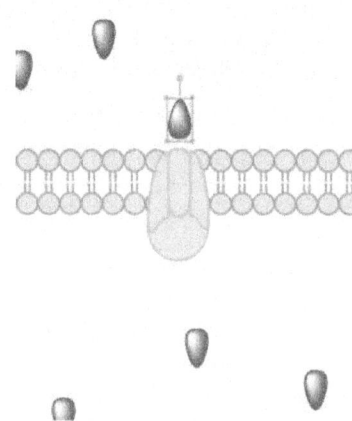

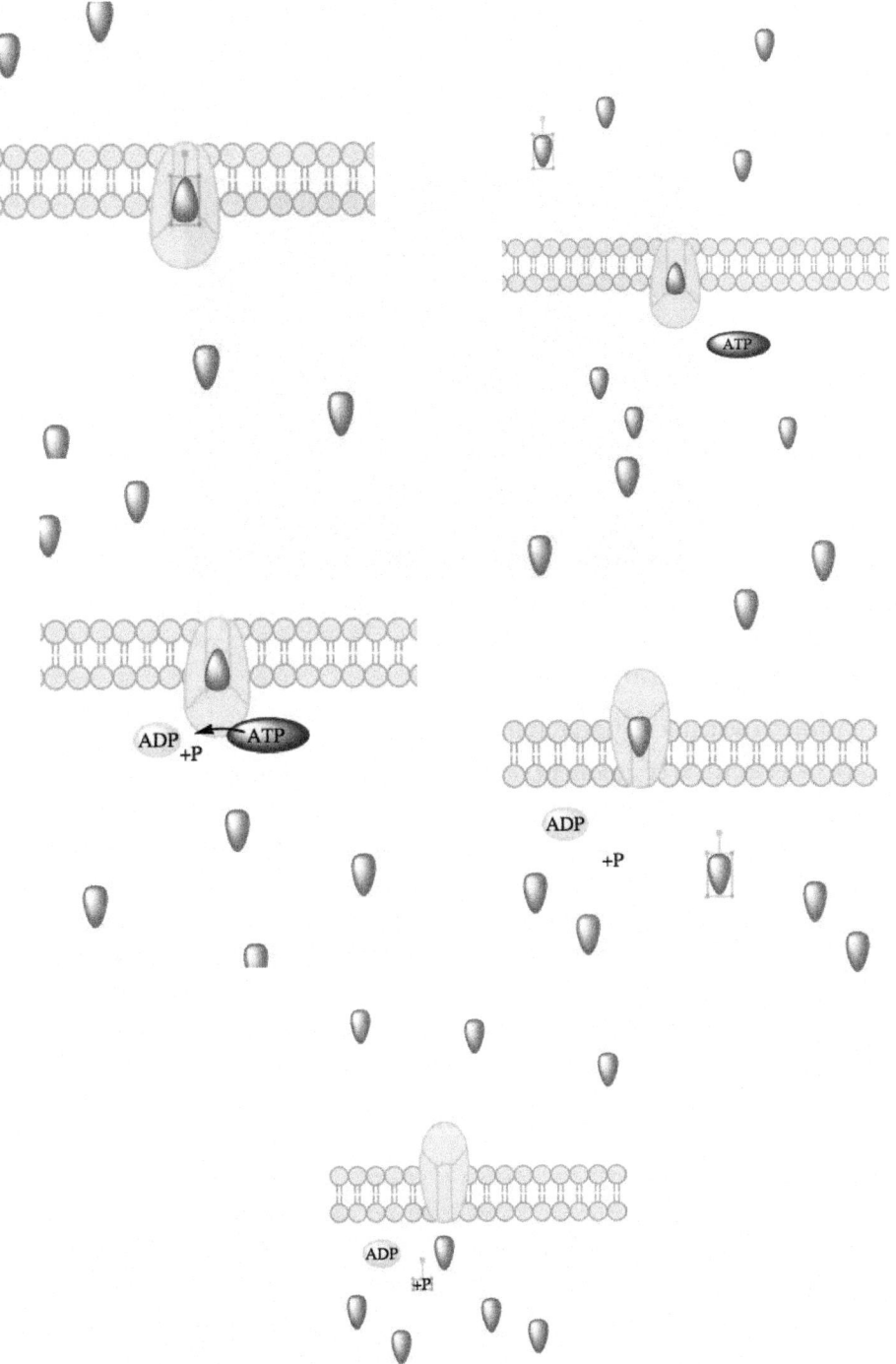

Fig. : Primary active transport, with the use of ATP, is depicted above progressing left to right and top to bottom.

The most universal example of ATP hydrolysis driving primary active transport in cells is the sodium-potassium pump. The sodium-potassium pump is responsible for controlling both sodium and potassium concentrations inside the cell. The sodium-potassium pump is extremely important in maintaining the cell's resting potential.

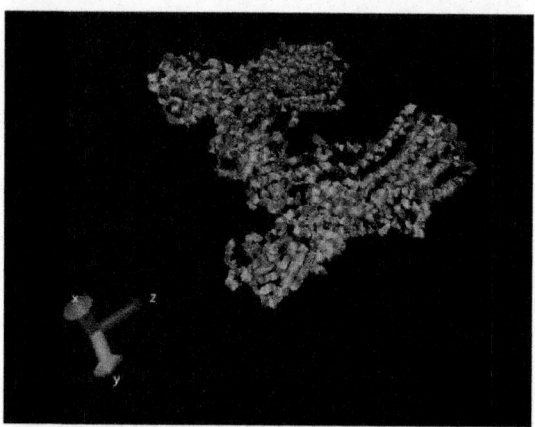

Fig. : Ribbon structure of a sodium-potassium ATPase pump.

Using Electrochemical Gradient Energy

An electrochemical gradient has two components:

1. An electrical component caused by charge difference on either side of the cellular membrane, and

2. A chemical component resulting from differing concentrations of ions across the cellular membrane.

The electrochemical gradient is generated by the presence of a proton (H^+) gradient. A proton gradient is an interconvertible form of energy that can ultimately be used by the transport protein to move particles across the cellular membrane.

A quintessential example of electrochemical gradient energy in primary active transport is the mitochondrial electron transport chain (ETC).The ETC uses the energy produced from the reduction of NADH to NAD^+ to create a proton gradient by pumping protons into the inner mitochondrial space.

Using Photon Energy

The energy stored in a photon, the basic unit of light, is used to generate a proton gradient through a process similar to that found in electrochemical gradients. The stepwise passing of electrons in an electron transport chain reduces a molecule like NADH and ultimately generates a proton gradient.

Plant photosynthesis is an example of primary active transport using photon energy. Chlorophyll absorbs a photon of light and consequently loses an electron which it passes pheophytin causing a subsequent electron transport chain. This

ETC ultimately ends in the reduction of NADH to NAD^+ creating a proton gradient across the chloroplast membrane.

Secondary Active Transport

Secondary active transport achieves an identical result as primary active transport in that particles are moved from low concentration to high concentration at the expense of energy. Secondary active transport, however, functions independent of direct ATP coupling. Rather, the electrochemical energy generated from pumping ions out of the cell is used. Secondary active transport is classified as either symporter of antiporter.

Symports

Symport secondary active transport uses a downhill movement of one particle to transport another particle against its concentration gradient. Symports move both particles in the same direction through a transmembrane transport protein.

A common symport example is SGLT1, a glucose symport. SGLT1 tranports one glucose molecule into the cell for every two sodium ions transported into the cell. The SGLT1 symport is located throughout the body, particularly in the nephron of the kidney.

Antiports

Antiport secondary active transport moves two or more different particles across the cellular membrane in opposite directions. Antiport secondary active transport moves one particle down its concentration gradient and uses the energy generated from that process to move another particle up its concentration gradient.

The sodium-calcium exchanger found throughout humans in excitable cells is a simple and common example of an antiport. Three sodium ions travel down their concentration gradient in exchange for one calcium ion.

Chapter 6

ANALYSIS OF A NOVEL METHOD FOR INHIBITING RUNAWAY REACTION *VIA* PROCESS MODELING

POLYMERIZATION REACTION AND POLYMER PROPERTIES

Polymerization

Chemical Reaction

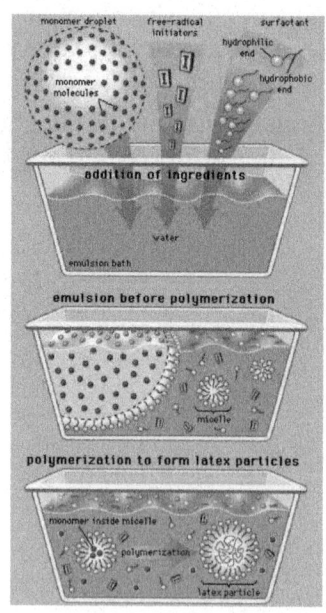

Polymerization, any process in which relatively small molecules, called mono-mers, combine chemically to produce a very large chainlike or network molecule, called a polymer. The monomer molecules may be all alike, or they may represent two, three, or more different compounds. Usually at least 100 monomer molecules must be combined to make a product that has certain unique physical properties — such as elasticity, high tensile strength, or the ability to form fibres — that differentiate polymers from substances composed of smaller and simpler molecules; often, many thousands of monomer units are incorporated in a single molecule of a polymer. The formation of stable covalent chemical bondsbetween the monomers sets polymerization apart from other processes, such as crystallization, in which large numbers of molecules aggregate under the influence of weak intermolecular forces.

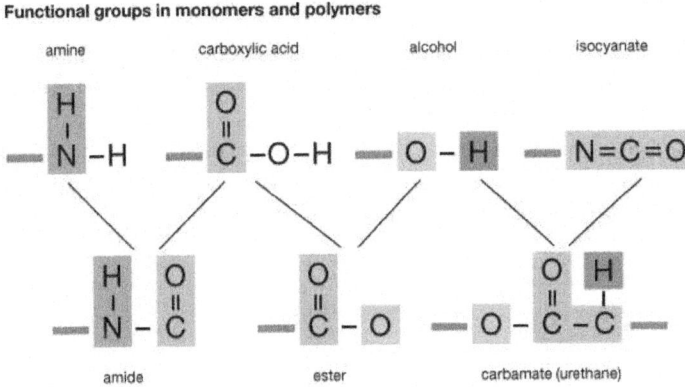

Two classes of polymerization usually are distinguished. In condensation polymerization, each step of the process is accompanied by the formation of a molecule of some simple compound, often water. Inaddition polymerization, monomers react to form a polymer without the formation of by-products. Addition polymerizations usually are carried out in the presence of catalysts, which in certain cases exert control over structural details that have important effects on the properties of the polymer.

Linear polymers, which are composed of chainlike molecules, may be viscous liquids or solids with varying degrees of crystallinity; a number of them can be dissolved in certain liquids, and they soften or melt upon heating. Cross-linked polymers, in which the molecular structure is a network, are thermosetting resins (*i.e.*, they form under the influence of heat but, once formed, do not melt or soften upon reheating) that do not dissolve in solvents. Both linear and cross-linked polymers can be made by either addition or condensation polymerization.

SCIENTIFIC PRINCIPLES

The field of polymers is so vast and the applications so varied, that it is important to understand how polymers are made and used. Since there are over 60,000 different plastics vying for a place in the market, knowledge of this important field can truly enrich our appreciation of this wonder material. Companies manufacture over 30 million tons of plastics each year, and spend large sums on research, development, and more efficient recycling methods. Below we learn some of the scientific principles involved in the production and processing of these fossil fuel derived materials known as polymers.

Polymerization Reactions

The chemical reaction in which high molecular mass molecules are formed from monomers is known as polymerization. There are two basic types of polymerization, chain-reaction (or addition) and step-reaction (or condensation) **polymerization.**

Chain-Reaction Polymerization

One of the most common types of polymer reactions is chain-reaction (addition) polymerization. This type of polymerization is a three step process involving two chemical entities. The first, known simply as a monomer, can be regarded as one link in a polymer chain. It initially exists as simple units. In nearly all cases, the monomers have at least one carbon-carbon double bond. Ethylene is one example of a monomer used to make a common polymer.

$$
\begin{array}{cc}
H & H \\
| & | \\
C & = C \quad \text{Ethylene} \\
| & | \\
H & H
\end{array}
$$

The other chemical reactant is a catalyst. In chain-reaction polymerization, the catalyst can be a free-radical peroxide added in relatively low concentrations. A free-radical is a chemical component that contains a free electron that forms a covalent bond with an electron on another molecule. The formation of a free radical from an organic peroxide is shown below:

$$
R-O-O-R \longrightarrow R-O^{\bullet} + R-O^{\bullet} \quad \begin{array}{l} \text{with } (\bullet) \\ \text{representing} \\ \text{the free electron} \end{array}
$$

In this chemical reaction, two free radicals have been formed from the one molecule of R_2O_2. Now that all the chemical components have been identified, we can begin to look at the polymerization process.

Step 1: Initiation

The first step in the chain-reaction polymerization process, initiation, occurs when the free-radical catalyst reacts with a double bonded carbon monomer, beginning the polymer chain. The double carbon bond breaks apart, the monomer bonds to the free radical, and the free electron is transferred to the outside carbon atom in this reaction.

$$
R-O^{\bullet} +
\begin{array}{cc}
H & H \\
| & | \\
C & = C \\
| & | \\
H & H
\end{array}
\longrightarrow
R-O-
\begin{array}{cc}
H & H \\
| & | \\
C & - C^{\bullet} \\
| & | \\
H & H
\end{array}
$$

Step 2: Propagation

The next step in the process, propagation, is a repetitive operation in which the physical chain of the polymer is formed. The double bond of successive monomers is opened up when the monomer is reacted to the reactive polymer chain. The free electron is successively passed down the line of the chain to the outside carbon atom.

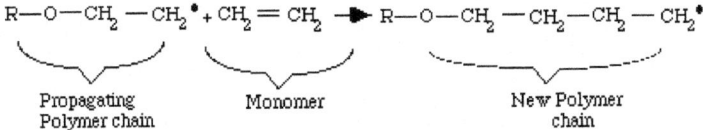

This reaction is able to occur continuously because the energy in the chemical system is lowered as the chain grows. Thermodynamically speaking, the sum of the energies of the polymer is less than the sum of the energies of the individual monomers. Simply put, the single bounds in the polymeric chain are more stable than the double bonds of the monomer.

Step 3: Termination

Termination occurs when another free radical (R-O·), left over from the original splitting of the organic peroxide, meets the end of the growing chain. This free-radical terminates the chain by linking with the last CH_2· component of the polymer chain. This reaction produces a complete polymer chain. Termination can also occur when two unfinished chains bond together. Both termination types are diagrammed below. Other types of termination are also possible.

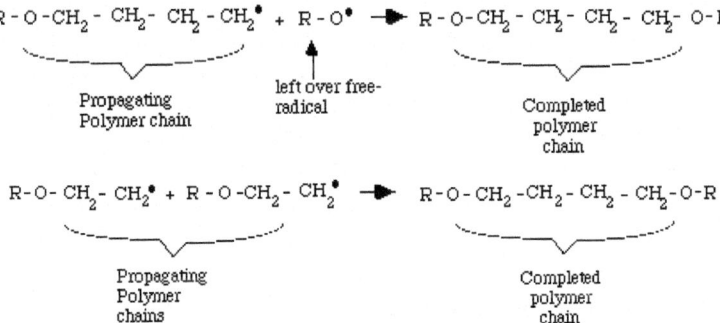

This exothermic reaction occurs extremely fast, forming individual chains of polyethylene often in less than 0.1 second. The polymers created have relatively high molecular weights. It is not unusual forbranches or cross-links with other chains to occur along the main chain.

Step-Reaction Polymerization

Step-reaction (condensation) polymerization is another common type of polymerization. This polymerization method typically produces polymers of lower molecular weight than chain reactions and requires higher temperatures to occur. Unlike addition polymerization, step-wise reactions involve two different types of di-functional monomers or end groups that react with one another, forming a chain. Condensation polymerization also produces a small molecular by-product (water, HCl, *etc.*). Below is an example of the formation of Nylon 66, a common polymeric clothing material, involving one each of two monomers, hexamethylene diamine and adipic acid, reacting to form a dimer of Nylon 66.

At this point, the polymer could grow in either direction by bonding to another molecule of hexamethylene diamine or adipic acid, or to another dimer. As the chain grows, the short chain molecules are called oligomers. This reaction process can, theoretically, continue until no further monomers and reactive end groups are available. The process, however, is relatively slow and can take up to several hours or days. Typically this process breeds linear chains that are strung out without any cross-linking or branching, unless a tri-functional monomer is added.

Polymer Chemical Structure

The monomers in a polymer can be arranged in a number of different ways. As indicated above, both addition and condensation polymers can be linear, branched, or cross-linked. Linear polymers are made up of one long continuous chain, without any excess appendages or attachments. Branched polymers have a chain structure that consists of one main chain of molecules with smaller molecular chains branching from it. A branched chain-structure tends to lower the degree of crystallinity and density of a polymer. Cross-linking in polymers occurs when primary valence bonds are formed between separate polymer chain molecules.

Chains with only one type of monomer are known as homopolymers. If two or more different type monomers are involved, the resulting copolymer can have several configurations or arrangements of the monomers along the chain. The four main configurations are depicted below:

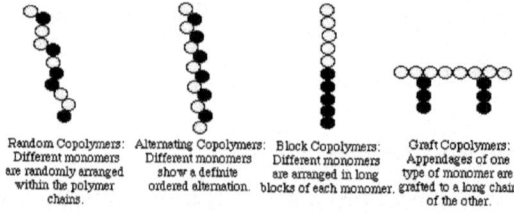

Random Copolymers: Alternating Copolymers: Block Copolymers: Graft Copolymers:
Different monomers   Different monomers   Different monomers Appendages of one
are randomly arranged show a definite   are arranged in long type of monomer are
within the polymer   ordered alternation.   blocks of each monomer. grafted to a long chain
chains.   of the other.

Fig. : Copolymer configurations.

Polymer Physical Structure

Segments of polymer molecules can exist in two distinct physical structures. They can be found in either crystalline or amorphous forms. Crystalline polymers are only possible if there is a regular chemical structure (*e.g.*, homopolymers or alternating copolymers), and the chains possess a highly ordered arrangement of their segments. Crystallinity in polymers is favored in symmetrical polymer chains, however, it is never 100%. These semi-crystalline polymers possess a rather typical liquefaction pathway, retaining their solid state until they reach their melting point at T_m.

Amorphous polymers do not show order. The molecular segments in amorphous polymers or the amorphous domains of semi-crystalline polymers are randomly arranged and entangled. Amorphous polymers do not have a definable T_m due to their randomness. At low temperatures, below their glass transition temperature (T_g), the segments are immobile and the sample is often brittle. As temperatures increase close to T_g, the molecular segments can begin to move. Above T_g, the mobility is sufficient (if no crystals are present) that the polymer can flow as a highly viscous liquid. The viscosity decreases with increasing temperature and decreasing molecular weight. There can also be an elastic response if the entanglements cannot align at the rate a force is applied (as in silly putty). This material is then described as visco-elastic. In a semi-crystalline polymer, molecular flow is prevented by the portions of the molecules in the crystals until the temperature is above T_m. At this point a visco-elastic material forms. These effects can most easily be seen on a specific volume versus temperature graph.

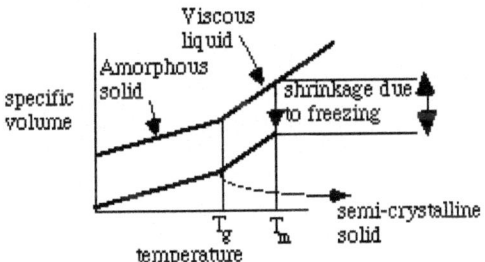

In the are between Tg and Tm, the semi-crystalline polymer is a tough solid. The amorphous material changes to a viscous liquid after Tg. This is when the material can be easily deformed.

Fig. : Specific Volume versus Temperature graph.

Members of the Polymer Family

Polymers can be separated into two different groups depending on their behavior when heated. Polymers with linear molecules are likely to be thermoplastic. These are substances that soften upon heating and can be remolded and recycled. They can be semi-crystalline or amorphous. The other group of polymers is known as thermosets. These are substances that do not soften under heat and pressure and cannot be remolded or recycled. They must be remachined, used as fillers, or incinerated to remove them from the environment.

Thermoplastics

Thermoplastics are generally carbon containing polymers synthesized by addition or condensation polymerization. This process forms strong covalent bonds within the chains and weaker secondary Van der Waals bonds between the chains. Usually, these secondary forces can be easily overcome by thermal energy, making thermoplastics moldable at high temperatures. Thermoplastics will also retain their newly reformed shape after cooling. A few common applications of thermoplastics include: parts for common household appliances, bottles, cable insulators, tape, blender and mixer bowls, medical syringes, mugs, textiles, packaging, and insulation.

Thermosets

Thermosets have the same Van der Waals bonds that thermoplastics do. They also have a stronger linkage to other chains. Strong covalent bonds chemically hold different chains together in a thermoset material. The chains may be directly bonded to each other or be bonded through other molecules. This "cross-linking" between the chains allows the material to resist softening upon heating. Thus, thermosets must be machined into a new shape if they are to be reused or they can serve as powdered fillers. Although thermosets are difficult to reform, they have many distinct advantages in engineering design applications including:

1. High thermal stability and insulating properties.
2. High rigidity and dimensional stability.
3. Resistance to creep and deformation under load.
4. Light-weight.

A few common applications for thermosets include epoxies (glues), automobile body parts, adhesives for plywood and particle board, and as a matrix for composites in boat hulls and tanks.

Polymer Processing

There are five basic processes to form polymer products or parts. These include; injection molding, compression molding, transfer molding, blow molding, and extrusion. Compression molding and transfer molding are used mainly for thermosetting plastics. Injection molding, extrusion and blow molding are used primarily with thermoplastics.

Injection Molding

This very common process for forming plastics involves four steps:

1. Powder or pelletized polymer is heated to the liquid state.
2. Under pressure, the liquid polymer is forced into a mold through an opening, called a sprue. Gates control the flow of material.

3. The pressurized material is held in the mold until it solidifies.

4. The mold is opened and the part removed by ejector pins.

Advantages of injection molding include rapid processing, little waste, and easy automation. Molded parts include combs, toothbrush bases, pails, pipe fittings, and model airplane parts.

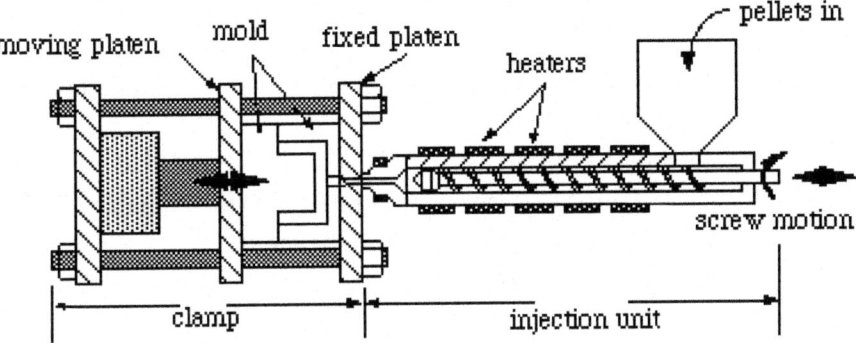

Fig. : Diagram of injection molding.

Compression Molding

This type of molding was among the first to be used to form plastics. It involves four steps:

1. Pre-formed blanks, powders or pellets are placed in the bottom section of a heated mold or die.

2. The other half of the mold is lowered and is pressure applied.

3. The material softens under heat and pressure, flowing to fill the mold. Excess is squeezed from the mold. If a thermoset, cross-linking occurs in the mold.

4. The mold is opened and the part is removed.

For thermoplastics, the mold is cooled before removal so the part will not lose its shape. Thermosets may be ejected while they are hot and after curing is complete. This process is slow, but the material moves only a short distance to the mold, and does not flow through gates or runners. Only one part is made from each mold.

Transfer Molding

This process is a modification of compression molding. It is used primarily to produce thermosetting plastics. Its steps are:

1. A partially polymerized material is placed in a heated chamber.

2. A plunger forces the flowing material into molds.

3. The material flows through sprues, runners and gates.

4. The temperature and pressure inside the mold are higher than in the heated chamber, which induces cross-linking.

5. The plastic cures, is hardened, the mold opened, and the part removed.

Mold costs are expensive and much scrap material collects in the sprues and runners, but complex parts of varying thickness can be accurately produced.

Blow Molding

Blow molding produces bottles, globe light fixtures, tubs, automobile gasoline tanks, and drums. It involves:

1. A softened plastic tube is extruded

2. The tube is clamped at one end and inflated to fill a mold.

3. Solid shell plastics are removed from the mold.

This process is rapid and relatively inexpensive.

Extrusion

This process makes parts of constant cross section like pipes and rods. Molten polymer goes through a die to produce a final shape. It involves four steps:

1. Pellets of the polymer are mixed with coloring and additives.

2. The material is heated to its proper plasticity.

3. The material is forced through a die.

4. The material is cooled.

An extruder has a hopper to feed the polymer and additives, a barrel with a continuous feed screw, a heating element, and a die holder. An adapter at the end of an extruder blowing air through an orifice into the hot polymer extruded through a ring die produces plastic bags and films.

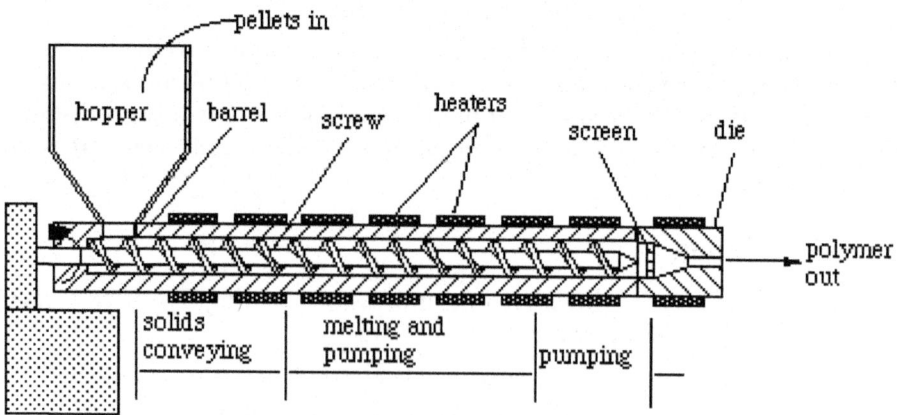

Fig. : Diagram of an extruder.

Table : Comparison of polymer processing techniques for
thermoplastics and thermosets.

Process	Thermoplastic (TP) or Thermoset (TS)	Advantages	Disadvantages
Injection Molding	TP, TS	It has the most precise control of shape and dimensions, is a highly automatic process, has fast cycle time, and the widest choice of materials.	It has high capital cost, is only good for large numbers of parts, and has large pressures in mold (20,000 psi).
Compression Molding	TS	It has lower mold pressures (1000 psi), does minimum damage to reinforcing fibers (in composites), and large parts are possible.	It requires more labor, longer cycle than injection molding, has less shape flexibility than injection molding, and each charge is loaded by hand.
Transfer Molding	TS	It is good for encapsulating metal parts and electronic circuits.	There is some scrap with every part and each charge is loaded by hand.
Blow Molding	TP	It can make hollow parts (especially bottles), stretching action improves mechanical properties, has a fast cycle, and is low labor.	It has no direct control over wall thickness, cannot mold small details with high precision, and requires a polymer with high melt strength.
Extrusion	TP	It is used for films, wraps, or long continuos parts (ie. pipes).	It must be cooled below its glass transition temperature to maintain stability.

RECYCLING: TODAY'S CHALLENGE, TOMORROW'S REWARD

Consumer waste in the United States poses a challenge to everyone. Waste solid materials can be grouped into the following categories:

- metals - aluminum, steel, *etc.*
- glass- clear, colored, *etc.*
- paper - newsprint, cardboard, *etc.*
- natural polymers- leather, grass, leaves, cotton, *etc.*
- synthetic polymers - synthetic rubbers, polyethylene terephthalate, polyvinyl chloride, *etc.*

Today, consumers are using more products and, therefore, producing more solid waste. As time goes by, we find ourselves with less space to put this waste. Eighty percent of all solid waste is buried in landfills. Today there are one third fewer landfills in operation than the 18,500 available a decade ago, making landfilling much more expensive.

Tipping fees, the charge to the waste hauler for dumping a load of solid waste, have been increasing regularly. Municipalities have imposed restrictions and/or have banned the startup of new landfills within their boundaries. As an example, 50% of New Jersey's solid waste is shipped out of state for landfill burial.

The amount which synthetic polymers contribute to the weight of solid waste will continue to go up as the use of plastics increases as projected below.

Table : America's plastic waste percentage by weight.

Year	Total Waste	Percentage Plastics
1960	76 million tons	2.7%
1984	133 million tons	7.2%
1995	142 million tons	8.4%
2000	159 million tons	9.8% (projected)

Plastics constitute between 14 and 22% of the volume of solid waste. One possible answer to this problem is recycling. In 1990, 1 to 2% of plastics, 29% of aluminum, 25% of paper, 7% of glass, and 3% of rubber and steel as post consumer wastes were recycled. Obviously, increasing the amount of plastics recycled would appear to be the answer. However, a major handicap in the reuse of plastics is that reprocessing adds a heat history, degrades properties and makes repeat use for the same application difficult. For example, the 58 gram, 2-liter polyethylene terephthalate (PET) beverage bottle consists of 48 g of PET, the rest being a high density polyethylene (HDPE) cup base, paper label, adhesive, and molded polypropylene (PP) cap. The cup base, label, adhesive and cap are contaminants in the recycling of the PET.

In response to the contaminants issue in plastic recycling, plastic products are being designed "reuse-friendly". Products are being made with recyclability as a viable means for disposal. At least one company has designed a 2-liter beverage bottle made of all PET for cost effective recycling. Concerning the reuse of recycled plastics, many organizations are reevaluating the use of recycled plastics. As an example, plastic beads are being used to remove paint from aircraft employing a "sand blasting" type method. Previously, harsh, environmentally unfriendly chemical solvents were used. The use of recycled plastics is only limited by the imagination of the designers and end users of the plastics.

Another reason for not discarding plastics is the conservation of energy. The energy value of polyethylene (PE) is 100 % of an equivalent mass of #2 heating oil. Polystyrene (PS) is 75%, while polyvinyl chloride (PVC) and PET are about 50%. With the energy value of a pound of #2 heating oil at 20,000 B.T.U., land filling plastics results in a waste of energy. Some countries, notably Japan, tap into the energy value of plastic and paper with waste-to-energy incinerators.

Another factor in the recycling equation is the economic trend of increasing tipping fees at landfills. In northeastern states, tipping fees have progressively increased, but in western states the fees have remained low due to the local government subsidies to landfills. As the cost of land filling of solid waste increases, so does the incentive to recycle. When the cost of land filling exceeds the cost of recycling, recycling will be a practical alternative to land filling.

These factors have led to certain recommendations by the United States Environmental Protection Agency. In order of highest to lowest, the EPA's recommendations are: source reduction, recycling, thermal reduction (incineration), and land filling. Each of these is not without its problems. Source reduction calls for the redesigning of packaging and/or the use of less, lighter, or more environmentally safe materials. The trade-off could mean reduced food packaging with the possibility of higher food spoilage rates. There would be fewer plastics, but more food in solid waste to be disposed. Whatever disposal method is chosen, the choice is complex. Whatever the costs, the consumer will bear them.

Recycling of Different Plastics

PET (Polyethylene Terephthalate)

In 1989, a billion pounds of virgin PET were used to make beverage bottles of which about 20% was recycled. Of the amount recycled, 50% was used for fiberfill and strapping. The reprocessors claim to make a high quality, 99% pure, granulated PET. It sells at 35 to 60% of virgin PET costs.

The major reuses of PET include sheet, fiber, film, and extrusions. When chemically treated, the recycled product can be converted into raw materials for the production of unsaturated polyester resins. If sufficient energy is used, the recycled product can be depolymerized to ethylene glycol and terephthalic acid and then repolymerized to virgin PET.

HDPE (High Density Polyethylene)

Of the plastics that have a potential for recycling, the rigid HDPE container is the one most likely to be found in a landfill. Less than 5% of HDPE containers are treated or processed in a manner that makes recycling easy. Virgin HDPE is used in opaque household and industrial containers used to package motor oil, detergent, milk, bleach, and agricultural chemicals.

There is a great potential for the use of recycled HDPE in base cups, drainage pipes, flower pots, plastic lumber, trash cans, automotive mud flaps, kitchen drain boards, beverage bottle crates, and pallets. Most recycled HDPE is a colored opaque material, that is available in a multitude of tints.

LDPE (Low Density Polyethylene)

LDPE is recycled by giant resin suppliers and merchant processors either by burning it as a fuel for energy or reusing it in trash bags. Recycling trash bags is a big business. Their color is not critical, therefore, regrinds go into black, brown, and to some lesser extent, green and yellow bags.

PVC (Polyvinyl Chloride)

There is much controversy concerning the recycling and reuse of PVC due to health and safety issues. When PVC is burned, the effects on the incinerator and quality of the air are often questioned. The Federal Food and Drug Administration (FDA) has ordered its staff to prepare environmental impact statements covering PVC's role in landfills and incineration. The burning of PVC releases toxic dioxins, furans, and hydrogen chloride. These fumes are carcinogenic, mutagenic, and teratagenic. This is one of the reasons why PVC must be identified and removed from any plastic waste to be recycled.

Currently, PVC is used in food and alcoholic beverage containers with FDA approval. The future of PVC rests in the hands of the plastics industry to resolve the issue of the toxic effects of the incineration of PVC. It is of interest to note that PVC accounts for less than 1% of land fill waste. When PVC is properly recycled, the problems of toxic emissions are minimized. Various recyclers have been able to reclaim PVC without the health problems. Uses for recycled PVC include aquarium tubing, drainage pipe, pipe fittings, floor tile, and nonfood bottles. When PVC is combined with other plastic waste it has been used to produce plastic lumber.

PS (Polystyrene)

PS and its manufacturers have been the target of environmentalists for several years. The manufacturers and recyclers are working hard to make recycling of PS as common as that of paper and metals. One company, Rubbermaid, is testing reclaimed PS in service trays and other utility items. Amoco, another large corporation, currently has a method that converts PS waste, including residual food, to an oil that can be re-refined.

The Future

Recycling is a viable alternative to all other means of dealing with consumer plastic waste. In response to the problem of mixed plastic waste, a coding system has been developed and adopted by the plastic industry. The code is a number and letter system. It applies to bottles exceeding 16 ounces and other containers exceeding 8 ounces. The number appears in the 3 bent arrow recycling symbol with the abbreviation of the plastic below the symbol.

Western European companies, especially the German firms Hoechst and Bayer, have entered the recyclable plastic market with success. With a high tech approach, they are devising new methods to separate and handle mixed plastics waste.

A potential use for recycled materials includes plastic lumber. The recycled plastic is mixed with wood fibers and processed into a replacement for lumber. The wood fibers would have become land fill if not reused. The end product is called Biopaste. This is expected to eventually become a multi-million dollar enterprise. Research and development continue to improve this product.

Recycling is a cost effective means of dealing with consumer plastic waste. Research to reduce the cost of recycling needs to continue. Recycling of plastics is not going to reach the level of the recycling programs of paper and some metals until lower cost, automatic methods of recycling are in place. Fortunately, the solutions to these problems are not beyond the scope of our technology or our minds.

Table : Major Plastic Resins and Their Uses.

Resin Code	Resin Name	Common Uses	Examples of Recycled Products
1	Polyethylene Terephthalate (PET or PETE)	Soft drink bottles, peanut butter jars, salad dressing bottles, mouth wash jars	Liquid soap bottles, strapping, fiberfill for winter coats, surfboards, paint brushes, fuzz on tennis balls, soft drink bottles, film
2	High density Polyethylene (HDPE)	Milk, water, and juice containers, grocery bags, toys, liquid detergent bottles	Soft drink based cups, flower pots, drain pipes, signs, stadium seats, trash cans, re-cycling bins, traffic barrier cones, golf bag liners, toys
3	Polyvinyl Chloride or Vinyl (PVC-V)	Clear food packaging, shampoo bottles	Floor mats, pipes, hoses, mud flaps
4	Low density Polyethylene (LDPE)	Bread bags, frozen food bags, grocery bags	Garbage can liners, grocery bags, multi purpose bags
5	Polypropylene (PP)	Ketchup bottles, yogurt containers, margarine, tubs, medicine bottles	Manhole steps, paint buckets, videocassette storage cases, ice scrapers, fast food trays, lawn mower wheels, automobile battery parts.
6	Polystyrene (PS)	Video cassette cases, compact disk jackets, coffee cups, cutlery, cafeteria trays, grocery store meat trays, fast-food sandwich container	License plate holders, golf course and septic tank drainage systems, desk top accessories, hanging files, food service trays, flower pots, trash cans

POLYMER SYNTHESIS

The study of polymer science begins with understanding the methods in which these materials are synthesized. Polymer synthesis is a complex procedure and can take place in a variety of ways. **Addition polymerization** describes the method where monomers are added one by one to an active site on the growing chain.

Addition Polymerization

The most common type of addition polymerization is free radical polymerization. A *free radical* is simply a molecule with an unpaired electron. The tendency for this free radical to gain an additional electron in order to form a pair makes it highly reactive so that it breaks the bond on another molecule by stealing an electron, learing that molecule with an unpaired election (which is another free radical). Free radicals are often created by the division of a molecule (known as an *initiator*) into two fragments along a single bond. The following diagram shows the formation of a radical from its initiator, in this case benzoyl peroxide.

The *stability* of a radical refers to the molecule's tendency to react with other compounds. An unstable radical will readily combine with many different molecules. However a stable radical will not easily interact with other chemical substances. The stability of free radicals can vary widely depending on the properties of the molecule. The *active center* is the location of the unpaired electron on the radical because this is where the reaction takes place. In free radical polymerization, the radical attacks one monomer, and the electron migrates to another part of the molecule. This newly formed radical attacks another monomer and the process is repeated. Thus the active center moves down the chain as the polymerization occurs.

There are three significant reactions that take place in addition polymerization: *initiation* (birth), *propagation* (growth), and *termination* (death).

Initiation Reaction

The first step in producing polymers by free radical polymerization is initiation. This step begins when an *initiator* decomposes into free radicals in the presence of monomers. The instability of carbon-carbon double bonds in the monomer makes them susceptible to reaction with the unpaired electrons in the radical. In this reaction, the active center of the radical "grabs" one of the electrons from the double bond of the monomer, leaving an unpaired electron to appear as a new active center at the end of the chain. Addition can occur at either end of the monomer. This process is illustrated in the following animation in which a chlorine atom possessing an unpaired electron (often indicated as cl-) initiates the reaction. As it collides with an ethylene molecule, it attracts one of the ethylene's pair of pi bonded electrons in forming a bond with one of the carbons. The other pi electron becomes the active center able to repeat this process with another ethylene molecule. The sigma bond between the carbons of the ethylene is not disturbed.

In a typical synthesis, between 60% and 100% of the free radicals undergo an initiation reaction with a monomer. The remaining radicals may join with each other or with an impurity instead of with a monomer. "Self destruction" of free radicals is a major hindrance to the initiation reaction. By controlling the monomer to radical ratio, this problem can be reduced.

Propagation Reaction

After a synthesis reaction has been initiated, the propagation reaction takes over. In the propagation stage, the process of electron transfer and consequent motion of the active center down the chain proceeds. In this diagram, (chain) refers to a chain of connected monomers, and X refers to a substituent group (a molecular fragment) specific to the monomer. For example, if X were a methyl group, the monomer would be propylene and the polymer, polypropylene.

$$
(chain)-CH_2\overset{\overset{\displaystyle H}{|}}{\underset{\underset{\displaystyle X}{|}}{C}} \cdot \; + \; \overset{\overset{\displaystyle H}{|}}{\underset{\underset{\displaystyle H}{|}}{C}}{=}\overset{\overset{\displaystyle H}{|}}{\underset{\underset{\displaystyle X}{|}}{C}} \longrightarrow (chain)-CH_2\overset{\overset{\displaystyle H}{|}}{\underset{\underset{\displaystyle X}{|}}{C}}-CH_2\overset{\overset{\displaystyle H}{|}}{\underset{\underset{\displaystyle X}{|}}{C}} \cdot \quad \cdots\cdot
$$

In free radical polymerization, the entire propagation reaction usually takes place within a fraction of a second. Thousands of monomers are added to the chain within this time. The entire process stops when the termination reaction occurs.

Termination Reaction

In theory, the propagation reaction could continue until the supply of monomers is exhausted. However, this outcome is very unlikely. Most often the growth of a polymer chain is halted by the termination reaction. Termination typically occurs in two ways: *combination* and *disproportionation*.

Combination occurs when the polymer's growth is stopped by free electrons from two growing chains that join and form a single chain. The following diagram depicts combination, with the symbol **(R)** representing the rest of the chain.

$$
(R)-CH_2\overset{\overset{\displaystyle H}{|}}{\underset{\underset{\displaystyle X}{|}}{C}} \cdot \; + \; \cdot\overset{\overset{\displaystyle H}{|}}{\underset{\underset{\displaystyle X}{|}}{C}}CH_2-(R) \longrightarrow (R)-CH_2\overset{\overset{\displaystyle H}{|}}{\underset{\underset{\displaystyle X}{|}}{C}}-\overset{\overset{\displaystyle H}{|}}{\underset{\underset{\displaystyle X}{|}}{C}}CH_2-(R)
$$

Disproportionation halts the propagation reaction when a free radical strips a hydrogen atom from an active chain. A carbon-carbon double bond takes the place of the missing hydrogen. Termination by disproportionation is shown in the diagram.

$$
(R)-CH_2\overset{\overset{\displaystyle H}{|}}{\underset{\underset{\displaystyle X}{|}}{C}} \cdot \; + \; \cdot\overset{\overset{\displaystyle H}{|}}{\underset{\underset{\displaystyle X}{|}}{C}}CH_2-(R) \longrightarrow (R)-CH_2\overset{\overset{\displaystyle H}{|}}{\underset{\underset{\displaystyle X}{|}}{C}}-H \; + \; \overset{\overset{\displaystyle H}{|}}{\underset{\underset{\displaystyle X}{|}}{C}}{=}CH-(R)
$$

Disproportionation can also occur when the radical reacts with an impurity. This is why it is so important that polymerization be carried out under very clean conditions.

Living Polymerization

There exists a type of addition polymerization that does not undergo a termination reaction. This so-called "living polymerization" continues until the monomer supply has been exhausted. When this happens, the free radicals become less active due to interactions with solvent molecules. If more monomers are added to the solution, the polymerization will resume.

Uniform molecular weights (low *polydispersity*) are characteristic of living polymerization. Because the supply of monomers is controlled, the chain length can be manipulated to serve the needs of a specific application. This assumes that the *initiator* is 100% efficient.

Statistical Analysis of Polymers

When dealing with millions of molecules in a tiny droplet, statistical methods must be employed to make generalizations about the characteristics of the polymer. It can be assumed in polymer synthesis, each chain reacts independently.

Therefore, the bulk polymer is characterized by a wide distribution of molecular weights and chain lengths. The *degree of polymerization* (DP) refers to the number of repeat units in the chain, and gives a measure of molecular weight. Many important properties of the final result are determined primarily from the distribution of lengths and the degree of polymerization. The following simulation allows you to examine the distribution of chain lengths under varying conditions.

STATISTICAL POLYMER GROWTH

In order to characterize the distribution of polymer lengths in a sample, two parameters are defined: *number average* and *weight average* molecular weight. The number average is just the sum of individual molecular weights divided by the number of polymers. The weight average is proportional to the square of the molecular weight. Therefore, the weight average is always larger than the number average. The following graph shows a typical distribution of polymers including the weight and number average molecular weights.

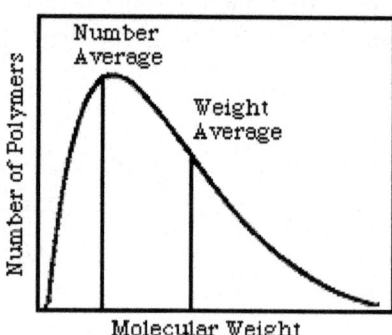

The molecular weight of a polymer can also be represented by the viscosity average molecular weight. This form of the molecular weight is found as a function of the viscosity of the polymer in solution (viscosity determines the rate at which the solution flows - the slower a solution moves, the more viscous it is said to be - and the polymer molecular weight influences the viscosity). The following simulation allows you to calculate the viscosity of a polymer solution, and use the data you find to produce the viscosity average molecular weight.

Viscosity Measurements

The degree of polymerization has a dramatic effect on the mechanical properties of a polymer. As chain length increases, mechanical properties such as ductility, tensile strength, and hardness rise sharply and eventually level off. This is schematically illustrated by the blue curve in the figure below.

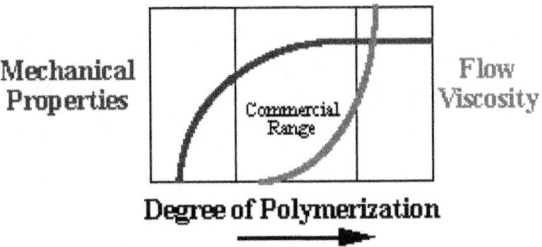

However, in polymer melts, for example, the flow viscosity at a given temperature rises rapidly with increasing DP for all polymers, as shown by the red curve in the diagram.

POLYMER STRUCTURE

Although the fundamental property of bulk polymers is the degree of polymerization, the physical structure of the chain is also an important factor that determines the macroscopic properties.

The terms **configuration** and **conformation** are used to describe the geometric structure of a polymer and are often confused. *Configuration* refers to the order that is determined by chemical bonds. The configuration of a polymer cannot be altered unless chemical bonds are broken and reformed. *Conformation* refers to order that arises from the rotation of molecules about the single bonds..

Configuration

The two types of polymer configurations are *cis* and *trans*. These structures can not be changed by physical means (*e.g.* rotation). The *cis* configuration arises when substituent groups are on the same side of a carbon-carbon double bond. *Trans* refers to the substituents on opposite sides of the double bond.

$$H_2C=CH_2 \quad cis \qquad trans$$

cis   trans

Stereoregularity is the term used to describe the configuration of polymer chains. Three distinct structures can be obtained. *Isotactic* is an arrangement where all substituents are on the same side of the polymer chain. A *syndiotactic* polymer chain is composed of alternating groups and *atactic* is a random combination of the groups. The following diagram shows two of the three *stereoisomers* of polymer chain.

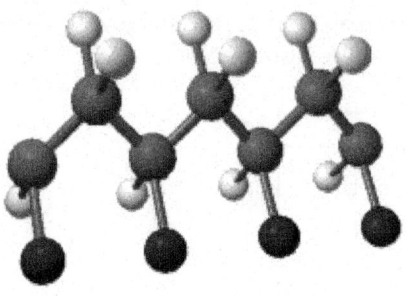

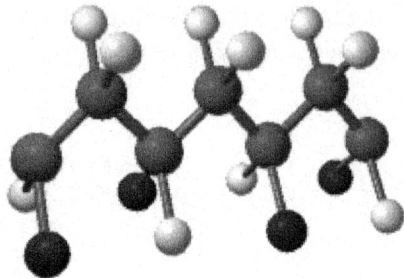

Isotactic   Syndiotactic

Conformation

If two atoms are joined by a single bond then rotation about that bond is possible since, unlike a double bond, it does not require breaking the bond.

The ability of an atom to rotate this way relative to the atoms which it joins is known as an adjustment of the *torsional* angle. If the two atoms have other atoms or groups attached to them then configurations which vary in torsional angle are known as *conformations*. Since different conformations represent varying distances between the atoms or groups rotating about the bond, and these distances determine the amount and type of interaction between adjacent atoms or groups, different conformation may represent different potential energies of the molecule. There several possible generalized conformations: Anti (Trans), Eclipsed (Cis), and Gauche (+ or -). The following animation illustrates the differences between them.

Conformation Lattice Simulation

Like the polymer growth simulation, the conformation lattice simulation takes a statistical approach to the study of polymers. Probabilities of the different conformations are assigned which produces a polymer chain with many possible shapes. Click the icon to enter the virtual laboratory.

POLYMER GROWTH LATTICE

Other Chain Structures

The geometric arrangement of the bonds is not the only way the structure of a polymer can vary. A *branched polymer* is formed when there are "side chains" attached to a main chain. A simple example of a branched polymer is shown in the following diagram.

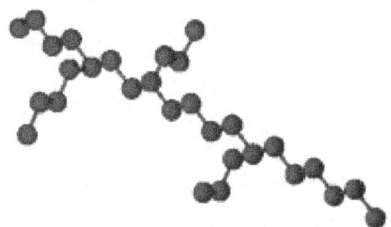

There are, however, many ways a branched polymer can be arranged. One of these types is called "*star-branching*". Star branching results when a polymerization starts with a single monomer and has branches radially outward from this point. Polymers with a high degree of branching are called *dendrimers* Often in these molecules, branches themselves have branches. This tends to give the molecule an overall spherical shape in three dimensions.

A separate kind of chain structure arises when more that one type of monomer is involved in the synthesis reaction. These polymers that incorporate more than one kind of monomer into their chain are called*copolymers*. There are three important types of copolymers. A *random copolymer* contains a random arrangement of the multiple monomers. A *block copolymer* contains blocks of monomers of the same type. Finally, a *graft copolymer* contains a main chain polymer consisting of one type of monomer with branches made up of other monomers. The following diagram displays the different types of copolymers.

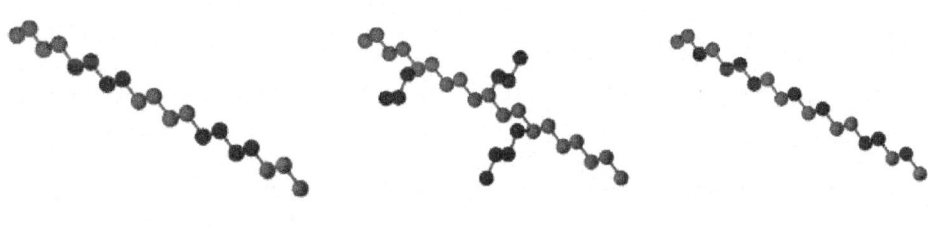

Block Copolymer   Graft Copolymer   Random Copolymer

An example of a common copolymer is Nylon. Nylon is an alternating copolymer with 2 monomers, a 6 carbon diacid and a 6 carbon diamine. The following picture shows one monomer of the diacid combined with one monomer of the diamine:

$$---\overset{\overset{\displaystyle O}{\|}}{C}-(CH_2)_4-\overset{\overset{\displaystyle O}{\|}}{C}-NH-(CH_2)_6-NH---$$

$$|\leftarrow\!\!-\text{Diacid}\!-\!\!\rightarrow|\leftarrow\!\!-\text{Diamine}\!-\!\!\rightarrow|$$

Cross-Linking

In addition to the bonds which hold monomers together in a polymer chain, many polymers form bonds between neighboring chains. These bonds can be formed directly between the neighboring chains, or two chains may bond to a third common molecule. Though not as strong or rigid as the bonds within the chain, these *cross-links* have an important effect on the polymer. Polymers with a high enough degree of cross-linking have "memory." When the polymer is stretched, the cross-links prevent the individual chains from sliding past each other. The chains may straighten out, but once the stress is removed they return to their original position and the object returns to its original shape.

One example of cross-linking is *vulcanization*. In vulcanization, a series of cross-links are introduced into an *elastomer* to give it strength. This technique is commonly used to strengthen rubber.

Classes of Polymers

Polymer science is a broad field that includes many types of materials which incorporate long chain structure of many repeat units.

*Elastomers,*or rubbery materials, have a loose cross-linked structure. This type of chain structure causes elastomers to possess memory. Typically, about 1 in 100 molecules are cross-linked on average. When the average number of cross-links rises to about 1 in 30 the material becomes more rigid and brittle. Natural and synthetic rubbers are both common examples of elastomers. *Plastics* are polymers which, under appropriate conditions of temperature and pressure, can be molded or shaped (such as blowing to form a film). In contrast to elastomers, plastics have a greater stiffness and lack reversible elasticity. All plastics are polymers but not all polymers are plastics. Cellulose is an example of a polymeric material which must be substantially modified before processing with the usual methods used for plastics. Some plastics, such as nylon and cellulose acetate, are formed into fibers (which are regarded by some as a separate class of polymers in spite of a considerable overlap with plastics). Some of the main chain polymer liquid crystals also are the constituents of important fibers. Every day plastics such as polyethylene and poly(vinyl chloride) have replaced traditional materials like paper and copper for a wide variety of applications.

POLYMER MORPHOLOGY

Molecular shape and the way molecules are arranged in a solid are important factors in determining the properties of polymers. From polymers that crumble to the touch to those used in bullet proof vests, the molecular structure, conformation and orientation of the polymers can have a major effect on the macroscopic properties of the material. The general concept of self-assembly enters into the organization of molecules on the micro and macroscopic scale as they aggregate into more ordered structures. Crystallization, is an example of the self-assembly process as is the orientational organization of liquid crystals.

Crystallinity

We need to distinguish here, between *crystalline* and *amorphous* materials and then show how these forms coexist in polymers. Consider a comparison between glass, an amorphous material, and ice which is crystalline.

The highly ordered crystalline structure of ice changes the apparent properties of the polarized light, and the ice appears bright. Glass and water, lacking that highly ordered structure, both appear dark.

The amorphous morphology of glass leads to very different properties from crystalline solids. This is illustrated in the heating process where the application of heat to glass turns it from a brittle solid-like material at room temperature to a viscous liquid. In contrast, the application of heat to ice turns it from solid to liquid. Crystalline melting leads to striking changes in optical properties during the melting process when observed through crossed polarizers. This is illustrated in the following movie of the melting of an organic crystalline material. Note that while the temperatures are not recorded, the entire process occurs over a very narrow temperature range.

The reasons for the differing behaviors lie mainly in the structure of the solids. Crystalline materials have their molecules arranged in repeating patterns. Table salt has one of the simplest atomic structures with its component atoms, Na^+ and Cl^-, arranged in alternating rows and the structure of a small cube. Salt, sugar, ice and most metals are crystalline materials. As such, they all tend to have highly ordered and regular structures. Amorphous materials, by contrast, have their molecules arranged randomly and in long chains which twist and curve around one-another, making large regions of highly structured morphology unlikely.

The morphology of most polymers is semi-crystalline. That is, they form mixtures of small crystals and amorphous material and melt over a range of temperature instead of at a single melting point. The crystalline material shows a high degree of order formed by folding and stacking of the polymer chains. The amorphous or glass-like structure shows no long range order.

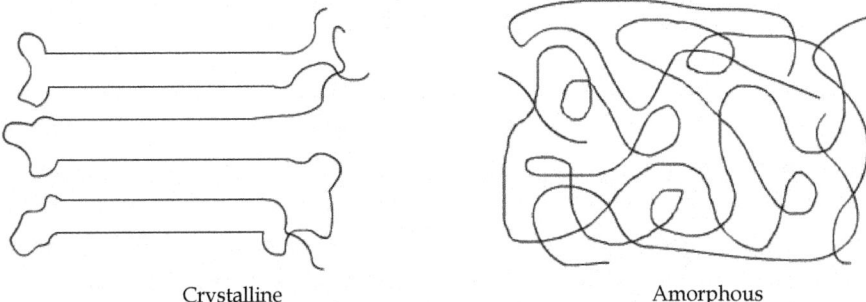

Crystalline   Amorphous

There are some polymers that are completely amorphous, but most are a combination with the tangled and disordered regions surrounding the crystalline areas. Such a combination is shown in the following diagram.

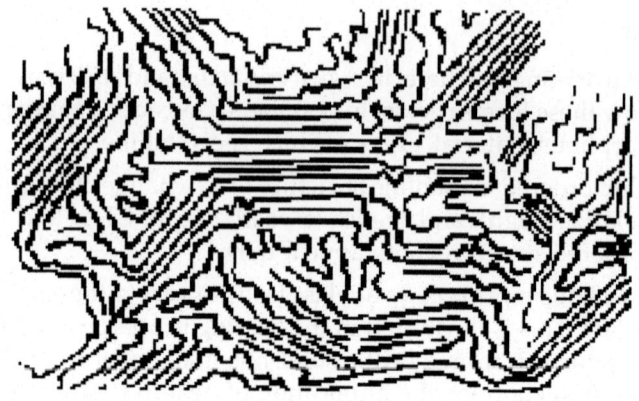

An *amorphous* solid is formed when the chains have little orientation throughout the bulk polymer. The *glass transition temperature* is the point at which the polymer hardens into an amorphous solid. This term is used because the amorphous solid has properties similar to glass.

In the crystallization process, it has been observed that relatively short chains organize themselves into crystalline structures more readily than longer molecules. Therefore, the *degree of polymerization* (DP) is an important factor in determining the crystallinity of a polymer. Polymers with a high DP have difficulty organizing into layers because they tend to become tangled.

The cooling rate also influences the amount of crystallinity. Slow cooling provides time for greater amounts of crystallization to occur. Fast rates, on the other hand, such as rapid quenches, yield highly amorphous materials. Subsequent annealing (heating and holding at an appropriate temperature below the crystalline melting point, followed by slow cooling) will produce a significant increase in crystallinity in most polymers, as well as relieving stresses.

Low molecular weight polymers (short chains) are generally weaker in strength. Although they are crystalline, only weak Van der Waals forces hold the lattice together. This allows the crystalline layers to slip past one another causing a break in the material. High DP (amorphous) polymers, however, have greater strength because the molecules become tangled between layers. For uses and examples of high and low DP polymers, see the section on Polymer Applications. In the case of fibers, stretching to 3 or more times their original length when in a semi-crystalline state produces increased chain alignment, crystallinity and strength.

In most polymers, the combination of crystalline and amorphous structures forms a material with advantageous properties of strength and stiffness.

Also influencing the polymer morphology is the size and shape of the monomers' substituent groups. If the monomers are large and irregular, it is difficult for the polymer chains to arrange themselves in an ordered manner, resulting in a more amorphous solid. Likewise, smaller monomers, and monomers that have a very regular structure (*e.g.* rod-like) will form more crystalline polymers.

THERMAL PROPERTIES OF POLYMERS

Polymer Glass Transition

In the study of polymers and their applications, it is important to understand the concept of the glass transition temperature, T_g. As the temperature of a polymer drops below T_g, it behaves in an increasingly brittle manner. As the temperature rises above the T_g, the polymer becomes more rubber-like. Thus, knowledge of T_g is essential in the selection of materials for various applications. In general, values of T_g well below room temperature define the domain of elastomers and values above room temperature define rigid, structural polymers.

This behavior can be understood in terms of the structure of glassy materials which are formed typically by substances containing long chains, networks of linked atoms or those that possess a complex molecular structure. Normally such materials have a high *viscosity* in the liquid state. When rapid cooling occurs to a temperature at which the crystalline state is expected to be the more stable, molecular movement is too sluggish or the geometry too awkward to take up a *crystalline* conformation. Therefore the random arrangement characteristic of the liquid persists down to temperatures at which the viscosity is so high that the material is considered to be solid. The term glassy has come to be synonymous with a persistent non-equilibrium state. In fact, a path to the state of lowest energy might not be available.

To become more quantitative about the characterization of the liquid-glass transition phenomenon and T_g, we note that in cooling an amorphous material from the liquid state, there is no abrupt change in volume such as occurs in the case of cooling of a crystalline material through its freezing point, T_f. Instead, at the glass transition temperature, T_g, there is a change in slope of the curve of specific volume vs. temperature, moving from a low value in the glassy state to a higher value in the rubbery state over a range of temperatures. This comparison between a crystalline material (1) and an amorphous material (2) is illustrated in the figure below. Note that the intersections of the two straight line segments of curve (2) defines the quantity T_g.

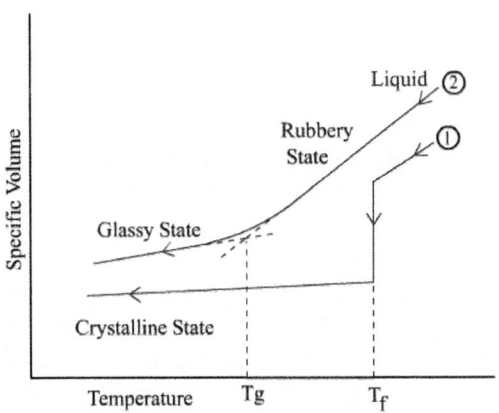

The specific volume measurements shown here, made on an amorphous polymer (2), are carried out in a dilatometer at a slow heating rate. In this apparatus, a sample is placed in a glass bulb and a confining liquid, usually mercury, is introduced into the bulb so that the liquid surrounds the sample and extends partway up a narrow bore glass capillary tube. A capillary tube is used so that relatively small changes in polymer volume caused by changing the temperature produce easily measured changes in the height of the mercury in the capillary.

The determination of T_g for amorphous materials, including polymers as mentioned above, by dilatometric methods (as well as by other methods) are found to be rate dependent. This is schematically illustrated in the figure below, again representing an amorphous polymer, where the higher value, T_{g2}, is obtained with a substantially higher cooling rate than for T_{g1}.

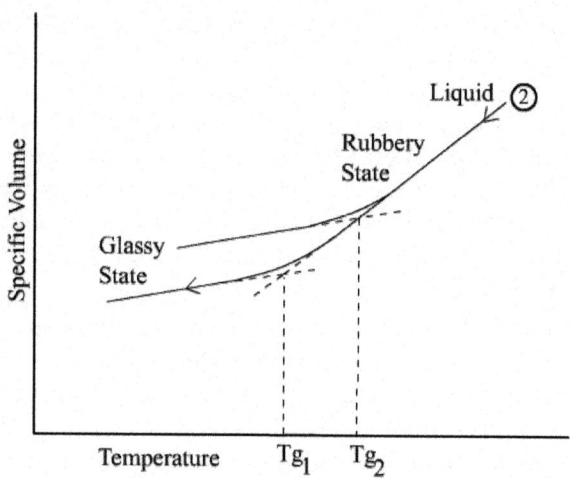

We can understand this rate dependence in terms of intermolecular relaxation processes. Since a glass is not an equilibrium phase, its properties will exhibit a time dependence, or physical aging. The primary portion of the relaxation behavior governing the glass transition in polymers can be related to their tangled chain structure where cooperative molecular motion is required for internal readjustments. At temperatures well above T_g, 10 to 50 repeat units of the polymer backbone are relatively free to move in cooperative thermal motion to provide conformational rearrangement of the backbone. Below T_g, the motion of these individual chains segments becomes frozen with only small scale molecular motion remaining, involving individual or small groups of atoms. Thus a rapid cooling rate or "quench" takes rubbery material into glassy behavior at higher temperatures (higher T_g).

While the dilatometer method is the more precise method of determining the glass transition temperature, it is a rather tedious experimental procedure and measurements of Tg are often made in a differential scanning calorimeter (DSC). In this instrument, the heat flow into or out of a small (10 – 20 mg) sample is measured as the sample is subjected to a programmed linear temperature change.

There are other methods of measurement such as density, dielectric constant and elastic modulus which are treated in texts on polymers. These methods are, of course, also rate dependent.

Tg and Mechanical Properties

Another important property of polymers, also strongly dependent on their temperatures, is their response to the application of a force, as indicated by two main types of behavior: *elastic* and *plastic*. Elastic materials will return to their original shape once the force is removed. Plastic materials will not regain their shape. In plastic materials, flow is occurring, much like a highly viscous liquid. Most materials demonstrate a combination of elastic and plastic behavior, showing plastic behavior after the elastic limit has been exceeded.

Glass is one of the few completely elastic materials while it is below its T_g. It will remain elastic until it reaches its breaking point. The T_g of glass occurs between 510 and 560 degrees C, meaning that it will always be a brittle solid at room temperature. In comparison, polyvinyl chloride (PVC) has a T_g of 83 degrees C, making it good, for example, for cold water pipes, but unsuitable for hot water. PVC also will always be a brittle solid at room temperature.

Adding a small amount of *plasticizer* to PVC can lower the T_g to – 40 degrees C. This addition renders the PVC a soft, flexible material at room temperature, ideal for applications such as garden hoses. A plasticized PVC hose can, however, become stiff and brittle in winter. In this case, as in any other, the relation of the T_g to the ambient temperature is what determines the choice of a given material in a particular application.

A striking example of the rate dependence of these viscoelastic properties is furnished by Silly Putty. Slowly pulling on two parts of the Silly Putty stretches it apart until it very slowly separates. Placing the Silly Putty on a table and hitting it with a hammer will shatter it.

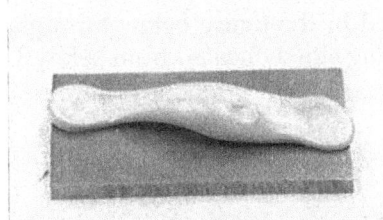
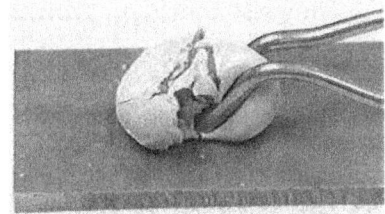

Slowly Deformed   Rapidly Deformed
Photos courtesy of Geon Corp.

The above images are representative of the behavior of a material above and below its glass transition temperature. The image on the (left) is Silly Putty that has been slowly stretched. The image on the (right) is Silly Putty which has been hit with a hammer. The speed of the hammer raised the rate of the application of the force and in turn raised the T_g. This caused the Silly Putty to react as if it were below its T_g and to shatter. Even though both reactions took place at the same ambient temperature, one reaction appeared to be above the effective T_g and the other appeared to be below.

Our focus has been on amorphous polymers in the preceding discussion but we have hardly touched on their mechanical properties. A further complication arises in dealing with general polymers from their semi-crystalline morphology in which amorphous regions and crystalline regions are intermingled. This gives rise to a mixed behavior depending on the percent crystallinity and on their temperature, relative to T_g of the amorphous regions. You are referred to texts on polymer science for basic discussion of these topic but the inhomogeneity of the material and its characteristics presents interesting analytical challenges.

Differential Scanning Calorimetry

In differential scanning calorimetry (DSC), the thermal properties of a sample are compared against a standard reference material which has no transition in the temperature range of interest, such as powdered alumina. Each is contained in a small holder within an adiabatic enclosure.

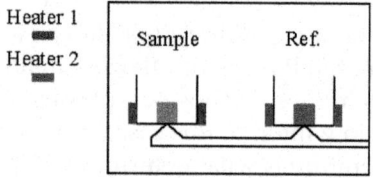

The temperature of each holder is monitored by a thermocouple and heat can be supplied electrically to each holder to keep the temperature of the two equal. A plot of the difference in energy supplied to the sample against the average temperature, as the latter is slowly increased through one or more thermal transitions of the sample yields important information about the transition, such as latent heat or a relatively abrupt change in heat capacity.

The glass transition process is illustrated in the figure below for a glassy polymer which does not crystallize and is being slowly heated from below T_g.

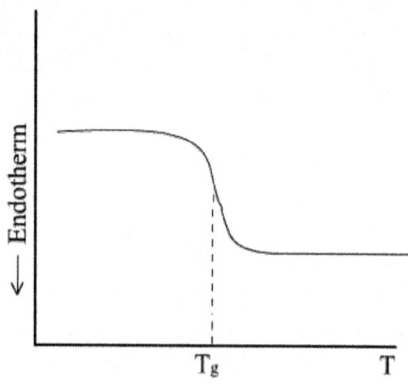

Here, the drop marked T_g at its midpoint represents the increase in energy supplied to the sample to maintain it at the same temperature as the reference material, due to the relatively rapid increase in the heat capacity of the sample as its temperature is raised through T_g. The addition of heat energy corresponds to this endothermal direction.

A melting process for the case of a highly crystalline polymer which is slowly heated through its melting temperature:

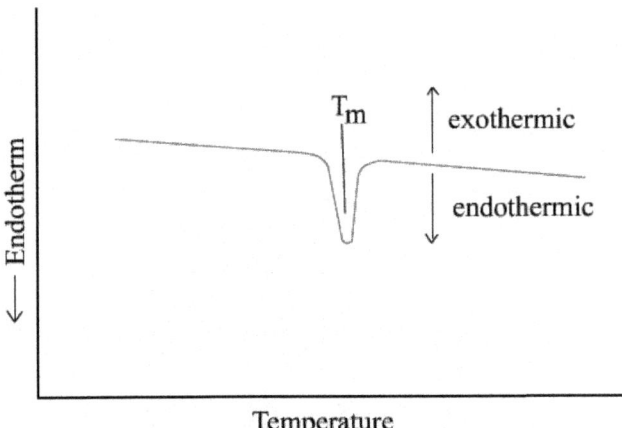

Again, as the melting temperature is reached, an endothermal peak appears because heat must be preferentially added to the sample to continue this essentially constant temperature process. The peak breadth is primarily related to the size and degree of perfection of the polymer crystals.

Note that if the process were reversed so that the sample were being cooled from the melt, the plot would be inverted. In that case, as both are being cooled by ambient conditions, even less heat would need to be supplied to the sample than to the reference material, in order that crystals can form. This corresponds to an exothermal process.

Use of the DSC on liquid crystals in connection with the identification of their phase transitions. An interesting exercise for the reader would be to predict the general form of a DSC plot for a semicrystalline polymer which has been rapidly quenched from the melt to a temperature below T_g. In the DSC plot, assume the temperature is slowly increased from this value below T_g to a value well above, thus allowing for significant increases in the chain mobility as temperatures above T_g are reached so that some crystallization can begin, well before the melting point is reached.

Applications of Polymers

Macromolecular science has had a major impact on the way we live. It is difficult to find an aspect of our lives that is not affected by polymers. Just 50 years ago, materials we now take for granted were non-existent. With further advances in the understanding of polymers, and with new applications being researched, there is no reason to believe that the revolution will stop any time soon.

Elastomers

Rubber is the most important of all *elastomers*. Natural rubber is a polymer whose repeating unit is isoprene. This material, obtained from the bark of the rubber tree, has been used by humans for many centuries. It was not until 1823, however, that rubber became the valuable material we know today. In that year, Charles Goodyear succeeded in "vulcanizing" natural rubber by heating it with sulfur. In this process, sulfur chain fragments attack the polymer chains and lead to *cross-linking*. The term vulcanization is often used now to describe the cross-linking of all elastomers.

Much of the rubber used in the United States today is a synthetic variety called styrene-butadiene rubber (SBR). Initial attempts to produce synthetic rubber revolved around isoprene because of its presence in natural rubber. Researchers eventually found success using butadiene and styrene with sodium metal as the *initiator*. This rubber was called Buna-S -- "Bu" from butadiene, "na" from the symbol for sodium, and "S" from styrene. During World War II, hundreds of thousands of tons of synthetic rubber were produced in government controlled factories. After the war, private industry took over and changed the name to styrene-butadiene rubber. Today, the United States consumes on the order of a million tons of SBR each year. Natural and other synthetic rubber materials are quite important.

Plastics

Americans consume approximately 60 billion pounds of plastics each year. The two main types of plastics are thermoplastics and thermosets. Thermoplastics soften on heating and harden on cooling while thermosets, on heating, flow and cross-link to form rigid material which does not soften on future heating. Thermoplastics account for the majority of commercial usage.

Among the most important and versatile of the hundreds of commercial plastics is polyethylene. Polyethylene is used in a wide variety of applications because, based on its structure, it can be produced in many different forms. The first type to be commercially exploited was called low density polyethylene (LDPE) or branched polyethylene. This polymer is characterized by a large degree of branching, forcing the molecules to be packed rather loosely forming a low density material. LDPE is soft and pliable and has applications ranging from plastic bags, containers, textiles, and electrical insulation, to coatings for packaging materials.

Another form of polyethylene differing from LDPE only in structure is high density polyethylene (HDPE) or linear polyethylene. This form demonstrates little or no branching, enabling the molecules to be tightly packed. HDPE is much more rigid than branched polyethylene and is used in applications where rigidity is important. Major uses of HDPE are plastic tubing, bottles, and bottle caps.

Other forms of this material include high and ultra-high molecular weight polyethylenes. HMW and UHMW, as they are known. These are used in applications where extremely tough and resilient materials are needed.

Fibers

Fibers represent a very important application of polymeric materials, including many examples from the categories of plastics and elastomers.

Natural fibers such as cotton, wool, and silk have been used by humans for many centuries. In 1885, artificial silk was patented and launched the modern fiber industry. Man-made fibers include materials such as nylon, polyester, rayon, and acrylic. The combination of strength, weight, and durability have made these materials very important in modern industry.

Generally speaking, fibers are at least 100 times longer than they are wide. Typical natural and artificial fibers can have *axial ratios* (ratio of length to diameter) of 3000 or more.

Synthetic polymers have been developed that posess desirable characteristics, such as a high softening point to allow for ironing, high *tensile strength*, adequate stiffness, and desirable fabric qualities. These polymers are then formed into fibers with various characteristics.

Nylon (a generic term for polyamides) was developed in the 1930's and used for parachutes in World War II. This synthetic fiber, known for its strength, elasticity, toughness, and resistance to abrasion, has commercial applications including clothing and carpeting. Nylon has special properties which distinguish it from other materials. One such property is the elasticity. Nylon is very elastic, however after elastic limit has been exceeded the material will not return to its original shape. Like other synthetic fibers, Nylon has a large electrical resistance. This is the cause for the build-up of static charges in some articles of clothing and carpets.

From textiles to bullet-proof vests, fibers have become very important in modern life. As the technology of fiber processing expands, new generations of strong and light weight materials will be produced.

Processing Polymers

Once a polymer with the right properties is produced, it must be manipulated into some useful shape or object. Various methods are used in industry to do this. **Injection molding** and **extrusion** are widely used to process plastics while **spinning** is the process used to produce fibers.

Injection Molding

One of the most widely used forms of plastic processing is injection molding. Basically, a plastic is heated above its glass transition temperature (enough so that it will flow) and then is forced under high pressure to fill the contents of a mold. The molten plastic in usually "squeezed" into the mold by a ram or a reciprocating screw. The plastic is allowed to cool and is then removed from the mold in its final form. The advantage of injection molding is speed; this process can be performed many times each second.

Extrusion

Extrusion is similar to injection molding except that the plastic is forced through a die rather than into a mold. However, the disadvantage of extrusion is that the objects made must have the same cross-sectional shape. Plastic tubing and hose is produced in this manner.

Spinning

The process of producing fibers is called spinning. There are three main types of spinning: melt, dry, and wet. Melt spinning is used for polymers that can be melted easily. Dry spinning involves dissolving the polymer into a solution that can be evaporated. Wet spinning is used when the solvent cannot be evaporated and must be removed by chemical means. All types of spinning use the same principle, so it is convenient to just describe just one. In melt spinning, a mass of polymer is heated until it will flow. The molten polymer is pumped to the face of a metal disk containing many small holes, called the spinneret. Tiny streams of polymer that emerge from these holes (called filaments) are wound together as they solidify, forming a long fiber. Speeds of up to 2500 feet/minute can be employed in spinning.

POLYMERIZATION REACTIONS

Free-Radical Polymerization Reactions

It isn't difficult to form addition polymers from monomers containing C=C double bonds; many of these compounds polymerize spontaneously unless polymerization is actively inhibited. One of the problems with early techniques for refining gasoline, for example, was the polymerization of alkene components when the gasoline was stored. Even with modern gasolines, deposits of "gunk" can form when a car or motorcycle is stored for extended periods of time without draining the carburetors.

The simplest way to catalyze the polymerization reaction that leads to an addition polymer is to add a source of a **free radical** to the monomer. The term *free radical* is used to describe a family of very reactive, short-lived components of a reaction that contain one or more unpaired electrons. In the presence of a free radical, addition polymers form by a chain-reaction mechanism that contains chain-initiation, chain-propagation, and chain- termination steps.

Chain Initiation

A source of free radicals is needed to initiate the chain reaction. These free radicals are usually produced by decomposing a peroxide such as di-*tert*-butyl peroxide or benzoyl peroxide. In the presence of either heat or light, these peroxides decompose to form a pair of free radicals that contain an unpaired electron.

Chain Propagation

The free radical produced in the chain-initiation step adds to an alkene to form a new free radical.

The product of this reaction can then add additional monomers in a chain reaction.

Chain Termination

Whenever pairs of radicals combine to form a covalent bond, the chain reactions carried by these radicals are terminated.

$$
CH_3-\underset{\underset{CH_3}{|}}{\overset{\overset{CH_3}{|}}{C}}-\ddot{O}-(CH_2CH_2)_n\cdot \ + \ \cdot(CH_2CH_2)_m-\ddot{O}-\underset{\underset{CH_3}{|}}{\overset{\overset{CH_3}{|}}{C}}-CH_3 \longrightarrow
$$

$$
CH_3-\underset{\underset{CH_3}{|}}{\overset{\overset{CH_3}{|}}{C}}-\ddot{O}-(CH_2CH_2)_{n+m}-\ddot{O}-\underset{\underset{CH_3}{|}}{\overset{\overset{CH_3}{|}}{C}}-CH_3
$$

The Formation of Branched Polymers

At first glance we might expect the product of the free-radical polymerization of ethylene to be a straight-chain polymer. As the chain grows, however, it begins to fold back on itself. This allows an *intramolecular* reaction to occur in which the site at which polymerization occurs is transferred from the end of the chain to a carbon atom along the backbone.

When this happens, branches are introduced onto the polymer chain. Free-radical polymerization of ethylene produces a polymer that contains branches on between 1 and 5% of the carbon atoms. Of these branches, 10% contain two carbon atoms, 50% contain four carbon atoms, and 40% are longer side chains.

Anionic Polymerization

Addition polymers can also be made by chain reactions that proceed through intermediates that carry either a negative or positive charge.

When the chain reaction is initiated and carried by negatively charged intermediates, the reaction is known as **anionic polymerization**. Like free-radical polymerizations, these chain reactions take place via chain-initiation, chain-propagation, and chain-termination steps.

The reaction is initiated by a Grignard reagent or alkyllithium reagent, which can be thought of a source of a negatively charged CH_3^- or $CH_3CH_2^-$ ion.

$$H-\underset{\underset{H}{|}}{\overset{\overset{H}{|}}{C}}-\underset{\underset{H}{|}}{\overset{\overset{H}{|}}{C}}-Li \quad \rightleftharpoons \quad H-\underset{\underset{H}{|}}{\overset{\overset{H}{|}}{C}}-\underset{\underset{H}{|}}{\overset{\overset{H}{|}}{C}}{:}^{-} \quad Li^{+}$$

The CH_3^- or $CH_3CH_2^-$ ion from one of these metal alkyls can attack an alkene to form a carbon-carbon bond.

$$H-\underset{\underset{H}{|}}{\overset{\overset{H}{|}}{C}}-\underset{\underset{H}{|}}{\overset{\overset{H}{|}}{C}}{:}^{-} \; + \; \overset{H}{\underset{H}{>}}C=C\overset{H}{\underset{X}{<}} \; \longrightarrow \; H-\underset{\underset{H}{|}}{\overset{\overset{H}{|}}{C}}-\underset{\underset{H}{|}}{\overset{\overset{H}{|}}{C}}-\underset{\underset{H}{|}}{\overset{\overset{H}{|}}{C}}-\underset{\underset{X}{|}}{\overset{\overset{H}{|}}{C}}{:}^{-}$$

The product of this chain-initiation reaction is a new carbanion that can attack another alkene in a chain-propagation step.

$$H-\underset{\underset{H}{|}}{\overset{\overset{H}{|}}{C}}-\underset{\underset{H}{|}}{\overset{\overset{H}{|}}{C}}-\underset{\underset{H}{|}}{\overset{\overset{H}{|}}{C}}-\underset{\underset{H}{|}}{\overset{\overset{H}{|}}{C}}{:}^{-} \; + \; \overset{H}{\underset{H}{>}}C=C\overset{H}{\underset{X}{<}} \; \longrightarrow H-\underset{\underset{H}{|}}{\overset{\overset{H}{|}}{C}}-\underset{\underset{H}{|}}{\overset{\overset{H}{|}}{C}}-\underset{\underset{H}{|}}{\overset{\overset{H}{|}}{C}}-\underset{\underset{X}{|}}{\overset{\overset{H}{|}}{C}}-\underset{\underset{H}{|}}{\overset{\overset{H}{|}}{C}}-\underset{\underset{X}{|}}{\overset{\overset{H}{|}}{C}}{:}^{-}$$

The chain reaction is terminated when the carbanion reacts with traces of water in the solvent in which the reaction is run.

$$CH_3CH_2(CH_2\underset{\underset{X}{|}}{CH})_n CH_2\underset{\underset{X}{|}}{CH}{:}^{-} \; + \; H_2O \; \longrightarrow \; CH_3CH_2(CH_2\underset{\underset{X}{|}}{CH})_n CH_2\underset{\underset{X}{|}}{CH_2} \; + \; OH^{-}$$

Cationic Polymerization

The intermediate that carries the chain reaction during polymerization can also be a positive ion, or cation. In this case, the **cationic polymerization** reaction is initiated by adding a strong acid to an alkene to form a carbocation.

$$CH_2{=}\underset{\underset{X}{|}}{CH} \; + \; H^{+} \; \longrightarrow \; CH_3\underset{\underset{X}{|}}{CH}^{+}$$

The ion produced in this reaction adds monomers to produce a growing polymer chain.

$$CH_3\underset{\underset{X}{|}}{CH}^{+} \; + \; CH_2{=}\underset{\underset{X}{|}}{CH} \; \longrightarrow \; CH_3\underset{\underset{X}{|}}{CH}CH_2\underset{\underset{X}{|}}{CH}^{+}$$

The chain reaction is terminated when the carbonium ion reacts with water that contaminates the solvent in which the polymerization is run.

$$CH_3\underset{\underset{X}{|}}{CH}(CH_2\underset{\underset{X}{|}}{CH})_n CH_2\underset{\underset{X}{|}}{CH}^{+} \; + \; H_2O \; \longrightarrow \; CH_3\underset{\underset{X}{|}}{CH}(CH_2\underset{\underset{X}{|}}{CH})_n \underset{\underset{X}{|}}{CH}OH \; + \; H^{+}$$

Advantages of Free-radical Versus Ionic Polymerization

The initiation step of ionic polymerization reactions has a much smaller activation energy than the equivalent step for free-radical polymerizations. As a result, ionic polymerization reactions are relatively insensitive to temperature, and can be run at temperatures as low as -70°C. Ionic polymerization therefore tends to produce a more regular polymer, with less branching along the backbone, and morecontrolled tacticity.

Because the intermediates involved in ionic polymerization reactions can't combine with one another, chain termination only occurs when the growing chain reacts with impurities or reagents that can be specifically added to control the rate of chain growth. It is therefore easier to control the average molecular weight of the product of ionic polymerization reactions.

Ionic polymerizations are more difficult to carry out on an industrial scale than free-radical polymerizations. Ionic polymerization is therefore only used for monomers that don't polymerize by the free-radical mechanism or to prepare polymers with a regular structure.

Coordination Polymerization

In 1963 Karl Ziegler and Giulio Natta received the Nobel prize in chemistry for their discovery of coordination compound catalysts for addition polymerization reactions. These **Ziegler-Natta catalysts**provide the opportunity to control both the linearity and tacticity of the polymer.

Free-radical polymerization of ethylene produces a low-density, branched polymer with side chains of one to five carbon atoms on up to 3% of the atoms along the polymer chain. Ziegler-Natta catalysts produce a more linear polymer, which is more rigid, with a higher density and a higher tensile strength. Polypropylene produced by free-radical reactions, for example, is a soft, rubbery, atactic polymer with no commercial value. Ziegler-Natta catalysts provide an isotactic polypropylene, which is harder, tougher, and more crystalline.

A typical Ziegler-Natta catalyst can be produced by mixing solutions of titanium(IV) chloride ($TiCl_4$) and triethylaluminum [$Al(CH_2CH_3)_3$] dissolved in a hydrocarbon solvent from which both oxygen and water have been rigorously excluded. The product of this reaction is an insoluble olive-colored complex in which the titanium has been reduced to the Ti(III) oxidation state.

The catalyst formed in this reaction can be described as *coordinately unsaturated* because there is an open coordination site on the titanium atom. This allows an alkene to act as a Lewis base toward the titanium atom, donating a pair of electrons to form a transition-metal complex.

The alkene is then inserted into a $Ti\text{-}CH_2CH_3$ bond to form a growing polymer chain and a site at which another alkene can bond.

Thus, the titanium atom provides a template on which a linear polymer with carefully controlled stereochemistry can grow.

Chain-Reaction Polymerization

Chain-reaction polymerization, sometimes called **addition polymerization**, requires an **initiator** to start the growth of the reaction. The largest family of polymers, vinyl polymers, are produced by chain polymerization reactions. A good example is the **free-radical polymerization** of styrene, which is initiated by a free radical (R) that reacts with styrene. The compound that is formed still is a free radical, which can react again.

This reaction eventually leads to the formation of polystyrene, a portion. Polystyrene prepared by free-radical polymerization is

Polystyrene can be represented using a shorthand notation.

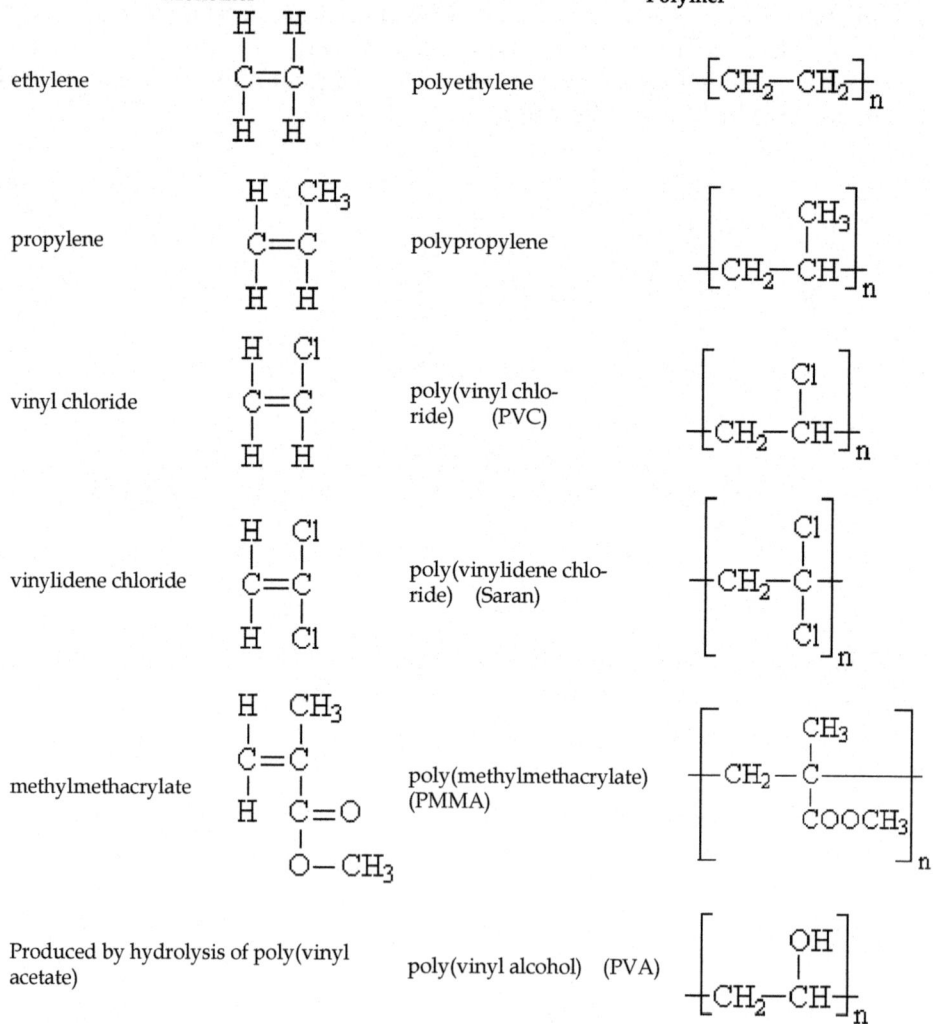

Styrene is an example of a vinyl monomer - a small molecules containing carbon-carbon double bonds. A few other important vinyl monomers. Some of these polymers are synthesized by methods involving initiators other than free radicals - carboanions, carbocations, or coordination compounds, for instance. Vinyl polymers are addition polymers, which have the same atoms as the monomer in their repeat units.

Step-Reaction Polymerization

In a step-reaction polymerization reaction, sometimes called **condensation polymerization**, the polymer chains grow by reactions that occur between two molecular species. An example is the polymerization reaction involving terephthalic acid and ethylene glycol, both of which are **bifunctional**.

$$HO-\overset{\overset{O}{\|}}{C}-\langle\bigcirc\rangle-\overset{\overset{O}{\|}}{C}-OH \quad + \quad HO-CH_2CH_2-OH$$

$$\downarrow -H_2O$$

Polymer formation begins with one diacid molecule reacting with one dialcohol molecule to eliminate a water molecule and form an ester. The ester unit has an alcohol on one end and acid on the other, which are available for further reactions.

$$HO-\overset{\overset{O}{\|}}{C}-\langle\bigcirc\rangle-\overset{\overset{O}{\|}}{C}-O-CH_2CH_2-OH$$

The eventual result is a **polyester** called **poly(ethylene terephthalate)** or more commonly, **PET.**

$$-O-\overset{C}{\underset{\overset{\|}{O}}{}}-\langle\bigcirc\rangle-\overset{C}{\underset{\overset{\|}{O}}{}}-O-CH^3CH^3-O-\overset{C}{\underset{\overset{\|}{O}}{}}-\langle\bigcirc\rangle-\overset{C}{\underset{\overset{\|}{O}}{}}-O-CH^3CH^3-O-\overset{C}{\underset{\overset{\|}{O}}{}}-\langle\bigcirc\rangle-\overset{C}{\underset{\overset{\|}{O}}{}}-O-CH^3CH^3-O-$$

PET is the plastic in soda bottles. It can be represented with a shorthand notation.

$$\left[O-CH_2CH_2-O-\overset{\overset{O}{\|}}{C}-\langle\bigcirc\rangle-\overset{\overset{O}{\|}}{C}\right]_n$$

Polyesters are **condensation polymers**, which contain fewer atoms within the polymer repeat unit than the reactants because of the formation of byproducts, such as H_2O or NH_3, during the polymerization reaction. Most synthetic fibers are condensation polymers.

A few types of condensation polymers. In the table, R and R' stand for organic groups.

Typical Monomers		Polymer Type
$HO-R-OH$ $Cl-\overset{\overset{O}{\|}}{C}-R'-\overset{\overset{O}{\|}}{C}-Cl$	polyester	$\left[R-O-\overset{\overset{O}{\|}}{C}-R'-\overset{\overset{O}{\|}}{C}-O\right]_n$

NH_2-R-NH_2 $HO-\overset{\overset{O}{\|}}{C}-R'-\overset{\overset{O}{\|}}{C}-OH$ polyamide (nylon) $\left[R-NH-\overset{\overset{O}{\|}}{C}-R'-\overset{\overset{O}{\|}}{C}-NH\right]_n$

$HO-R-OH$ $O=C=N-R'-N=C=O$ polyurethane $\left[NH-\overset{\overset{O}{\|}}{C}-O-R-O-\overset{\overset{O}{\|}}{C}-NH-R'\right]_n$

$HO-R-OH$ $Cl-\overset{\overset{O}{\|}}{C}-Cl$ polycarbonate $\left[O-R-O-\overset{\overset{O}{\|}}{C}\right]_n$

KINETIC ANALYSIS OF THE FREE-RADICAL POLYMERIZATION OF STYRENE

Polystyrene is a major commodity polymer used for a wide variety of commercial applications, and a vast majority of this polymer is synthesized via free-radical polymerization. The polymerization of vinyl monomers like styrene is often initiated by the thermal generation of radical species. For example, benzoyl peroxide thermally decomposes to form benzoyl radicals that initiate the polymerization of styrene to polystyrene (Scheme 1).

In principle, the rate of propagation in radical polymerization can be controlled by changing the initiator concentration. Mathematically, this is illustrated by the first-order rate equation for radical propagation:

$$R_p = -\frac{d[M]}{dt} = k_p \left(\frac{k_d f[I]}{k_t}\right)^{1/2} [M].$$

Here, f is the efficiency of an initiator I, k_d is the rate constant of initiator decomposition, k_p is the rate constant for propagation for a monomer M, and k_t

is the rate constant for termination. In this experiment, we have examined the validity of this rate equation for the bulk, free-radical polymerization of styrene. In particular, we have used Equation 1 to verify that propagation is first-order in styrene concentration, and to calculate the propagation rate constant k_p for the polymerization.

Experimental Section

Styrene polymerizations were performed in 6-inch test tubes, to which styrene and benzoyl peroxide were added in the amounts. The mixture was agitated at room temperature until the peroxide was completely dissolved. The test tube was then placed in a warm water bath (73.5 °C) and the polymerization was allowed to proceed until the solution viscosity began to increase noticeably (as determined by the behavior of bubbles in the tube). The mixture was then removed from the bath and the polymer was precipitated from cold hexanes (50 mL) to yield white, gooey material. The polymer was collected and redissolved in methyl ethyl ketone (2 mL), and reprecipitated from hexanes (30 mL). The polymer was recovered and placed in an aluminum pan that had been treated with a release agent. Residual solvent and monomer was removed by drying the polymer in a vacuum oven at 70 °C for one week, after which the isolated yield was recorded.

Table. Initiator concentration and monomer conversion in free-radical polymerization of styrene.

BP used (g)	BP used (mmol)	$[BP]_o$ (M)	reaction time (sec)	PS produced (g)	S con-sumed (%)	S con-sumed (mmol)	[S] con-sumed (M)	R_p (M/ sec x 10⁴)
0.108	0.44	0.089	1500	0.181	3.77	1.74	0.329	2.32
0.225	1.05	0.211	1320	0.341	7.51	3.28	0.655	4.96
0.424	1.75	0.350	1320	0.415	9.14	3.98	0.797	6.04
0.55	2.27	0.454	960	0.362	7.97	3.48	0.695	7.24
0.617	2.55	0.510	960	0.393	8.66	3.77	0.755	7.86
0.853	3.52	0.704	1020	0.405	8.92	3.89	0.778	7.63

Results and Discussion

At low conversions of monomer (less than about 10%) it can safely be assumed that the concentrations of monomer and initiator remain constant throughout the polymerization, such that

$[M] = [M]_0$ and $[I] = [I]_0$.

In this case, Equation can be rewritten as:

$$R_p = -\frac{\Delta[M]}{\Delta t} = k_p [M]_o [I]_o^{1/2} \left(\frac{k_d f}{k_t} \right)^{1/2}$$

This relationship should hold only if propagation is first-order in monomer concentration. One goal of this experimental study, as a result, was to test the relationship between R_p and [I]. A plot of

$$\Delta[S]/\Delta t \text{ vs. } [BP]_0^{1/2}$$

should have slope $k_p[M]_0(k_d f/k_t)^{1/2}$, which will yield the polymerization rate constant k_p.

Propagation rates were calculated from the molar conversion of styrene with respect to polymerization time at six different initiator concentrations. All of the polymerization reactions advanced to levels of conversion (3-10%) sufficient for rate determination using the method of initial rates. The propagation rates calculated from these experiments were observed to consistently increase with increasing initiator concentration.

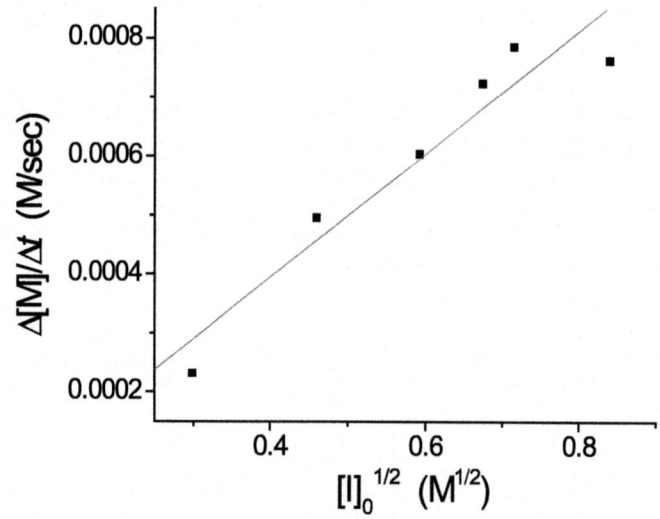

Fig. : Plot of propagation rate R_p with respect to root initiator concentration $[I]^{1/2}$ for the bulk polymerization of styrene at low monomer conversion.

Figure shows a plot of R_p versus $[I]^{1/2}$ for the benzoyl peroxide-initiated polymerization of styrene; linear regression of this data yielded a slope of 0.011 $M^{1/2}$/sec. Given $k_d = 1.0 \times 10^{-5}$ /sec, $k_t = 6 \times 10^7$ /M·sec, and $f = 0.8$, the calculated $k_p = 350$ /sec. This is consistent with published values of k_p for styrene between 50-400 /sec. The correlation coefficient (R^2) for this linear regression is 0.92, and several of the individual data points deviate significantly from the best-fit regression line. Errors in the data are expected to arise from difficulties in the recovery of the polystyrene after precipitation. Incomplete recovery of the small amounts of polystyrene prepared in these experiments would lead to inaccurately low propagation rates, whereas an inability to quantitatively remove residual monomer or solvent would lead to inaccurately high propagation rates. These data suggest that the recovered polymer may have contained residual materials other than polystyrene that were not removed by treatment of the sample in the vacuum oven.

SUSPENSION POLYMERIZATION

Hoffman and Delbruch first developed suspension polymerization in 1909. In suspension polymerization the initiator is soluble in the monomer phase, which is dispersed by comminuting into the dispersion medium (usually water) to form droplets. The solubility of the dispersed monomer (droplet) phase and also the resultant polymer in the dispersion medium are usually low. The volume fraction of the monomer phase is usually within the range 0. 1-0. 5. Polymerization reactions may be performed at lower monomer volume fractions, but are not usually economically viable. At higher volume fractions, the concentration of continuous phase may be insufficient to fill the space between droplets. Polymerization proceeds in the droplet phase, and in most cases occurs by a free radical mechanism. Suspension polymerization usually requires the addition of small amounts of a stabilizer to hinder coalescence and break-up of droplets during polymerization.

The size distribution of the initial emulsion droplets and, hence, also of the polymer beads that are formed, is dependent upon the balance between droplet break-up and droplet coalescence. This is in turn controlled by the type and speed of agitator used, the volume fraction of the monomer phase, and the type and concentration of stabilizer used. If the polymer is soluble in the monomer, a gel is formed within the droplets at low conversion leading to harder spheres at high conversion. If the polymer is insoluble in the monomer solution, precipitation will occur within the droplets, which will result in the formation of opaque, often irregularly shaped particles. If the polymer is partially soluble in the monomer mixture, the composition of the final product can be difficult to predict. Polymer beads find applications in a number of technologies, such as molding plastics. However, their largest application is in chromatographic separation media (as ion exchange resin and as supports for enzyme immobilization). Such applications frequently require large particle surface areas, which necessitates the formation of pores (of the required dimensions) in the bead structure.

The polymer beads may be made porous by the inclusion of an inert diluent (or porogen) to the monomer phase, which may be extracted after polymerization. Other additions to the monomer phase can include UV stabilizers (aromatic ketones and esters), heat stabilizers (ethylene oxide derivatives and inorganic metal salts), molding lubricants and foaming agents (porogens).

Polymeric Stabilizers

Typical polymeric stabilizers used for oil-in-water suspension polymerization reactions are poly (vinyl alcohol) -co- (vinyl acetate) (formed from the partial hydrolysis (80-90%) of polyvinyl acetate), poly (vinyl-pyrrolidone), salts of acrylic acid polymers, cellulose ethers and natural gums.

Polymeric stabilizers used in inverse suspension polymerization reactions include block copolymers poly (hydroxy-stearic acid) -co-poly) ethylene oxide). Surfactants used for oil-in-water suspensions include spans and the anionic emulsifier (sodium 12-butinoyloxy-9-octadecenate).

Polymerization Conditions and Kinetics

Extensive studies have shown that, in general, reaction kinetics in suspension polymerization is found to show good agreement with bulk phase kinetics (in absence of any monomer diluent). This observation suggests that in suspension polymerization, the emulsification conditions (agitation conditions, emulsion droplet size and concentration / type of stabilizer) appear to have little effect on reaction kinetics. Moreover, it can be concluded that any mass transfer between two phases in the emulsion does not affect the overall reaction rate. The major challenge in designing a suspension reaction is therefore the formation of a stable emulsion, preferably having a uniform size distribution. The monomer droplets are large enough to contain a large enough to contain a large number of free radicals (may be as many as 10^s) and this is why the polymerization in general proceeds with a similar mechanism to that of bulk polymerization, particularly when the polymer is soluble in the monomer.

Experiment

Styrene and 1, 4-divinylbenzene (the latter as 50-60% solution in ethyl benzene) are destabilized and distilled.

A three-necked flask, fitted with stirrer (preferably with revolution counter), thermometer, reflux condenser and nitrogen inlet, is evacuated and filled with nitrogen three times. 250 mg of poly (vinyl alcohol) are placed in the flask and dissolved in 150 mL of de-aerated water at 50°C. A freshly prepared solution of 0.25g (1.03 mmol) of dibenzoyl peroxide in 25 mL (0.22 mmol) of styrene and 2 mL (7 mmol) of 1, 4-divinylbenzene is added with constant stirring so as to produce an emulsion of fine droplets of monomer in water. This is heated to 90°C on a water bath while maintaining a constant rate of stirring and passing a gentle stream of nitrogen through the reaction vessel. After about 1 h (about 5% conversion) the cross-linking becomes noticeable (gelation). Stirring is continued for another 7 h at 90° C, the reaction mixture then being allowed to cool to room temperature while stirring. The supernatant liquid is decanted from the beads, which are washed several times with methanol and finally stirred for another 2 h with 200 ml of methanol. The polymer is filtered off and dried overnight in vacuum at 50° C. Yield: practically quantitative.

FEATURES OF CONTROLLED/"LIVING" RADICAL POLYMERIZATIONS

The introductory slide from the home page, reproduced below, indicates that Controlled/"Living" Radical Polymerization (CRP) procedures can be used for the preparation of copolymers, incorporating a broad spectrum of radically (co)polymerizable monomers forming materials with predetermined molecular weight, and narrow molecular weight distribution. The most recent work on conducting an ATRP with low concentration of transition metals indicates that some control over the breadth of the MWD is also possible. Here, and elsewhere in the text the word "control" and/or "controlled" means that if the polymerization process conditions are selected so that the contributions of the chain breaking processes are insignificant compared to chain propagation, then synthesis of polymers with predetermined molecular weights, low polydispersity and site specific functionalities become a reality.

What Can Controlled/Living Polymerizations Do?

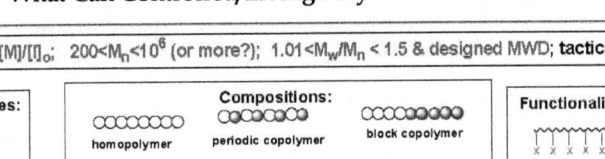

$$DP_n = \Delta[M]/[I]_0; \quad 200 < M_n < 10^6 \text{ (or more?)}; \quad 1.01 < M_w/M_n < 1.5 \text{ \& designed MWD; tactility}$$

Topologies: linear, star, comb / brush, network/ crosslinked, dendritic / hyperbranched

Compositions: homopolymer, periodic copolymer, block copolymer, random copolymer, tapered / gradient copolymer, graft copolymer

Monomers:
Vinyl: (Me), R
-Ar (Sty), Vi (Bu & IP), Pyr, CN, Cl, Br, Me, H

Systems: bulk, solution (org., H_2O, CO_2), suspension, emulsion, ...

Transformations: Hybrids: Z-N synthetic/natural, ROMP organic/inorganic, PCond. surfaces; IPN; etc.

Functionalities: side-functional groups, end-functional polymers, telechelic polymers, site-specific functional polymers, macromonomers, multifunctional

Carnegie Mellon University

FEATURE 1 First-order kinetic behavior:

i.e. the polymerization rate (R_p) with respect to the log of the monomer concentration ([M]) is a linear function of time. This is due to the negligible contribution of non-reversible termination, so that the concentration of the active propagating species ([P*]) is constant.

$$R_p = \frac{-d[M]}{dt} = k_p[P^*][M] \tag{CRP. 1}$$

$$\ln\frac{[M]_0}{[M]} = k_p[P^*]t = k_p^{app}[P^*]t \text{ (if } [P^*] \text{ is constant)} \tag{CRP.2}$$

k_p is the propagation constant.

The consequence of equation CRP.2 and the effect of changes in *P** are illustrated in below:

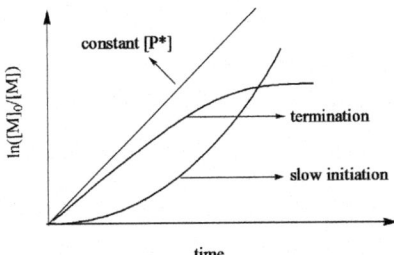

Illustration of FEATURE 1, the depence of Ln([M]0/[M]) on time

This semilogarithmic plot is very sensitive to any change of the concentration of the active propagating species. A constant $[P^*]$ is revealed by a straight line. A steady $[P^*]$ in a CRP is established by balancing the rates of activation and deactivation and *not* by balancing the rates of initiation and termination as in a conventional radical polymerization. An upward curvature in the kinetic plot indicates an increase in $[P^*]$, which occurs in case of slow initiation. On the other hand, a downward curvature suggests a decrease in $[P^*]$, which may result from termination reactions increasing the concentration of the persistent radical, or some other side reactions such as the catalytic system being poisoned or redox processes on the radical.

It should also be noted that the semilogarithmic plot is not sensitive to chain transfer processes or slow exchange between different active species, since they do not affect the number of the active propagating species.

FEATURE 2 Pre-determinable degree of polymerization:

(DP_n), *i.e.* the number average molecular weight (M_n) is a linear function of monomer conversion.

$$DP_n = \frac{M_n}{M_0} = \frac{\Delta[M]}{[I]_0} = \frac{[M]_0}{[I]_0} \ (conversion) \qquad \text{(CRP.3)}$$

This result comes from maintaining a constant number of chains throughout the polymerization, which requires the following two conditions:

1. Initiation should be sufficiently fast so that essentially all chains are propagating before the reaction is stopped; and
2. An absence of chain transfer reactions that increases the total number of chains.

The following figure illustrates feature 2 and shows the ideal growth of molecular weight with conversion, as well as the effects of slow initiation and chain transfer on the molecular weight evolution.

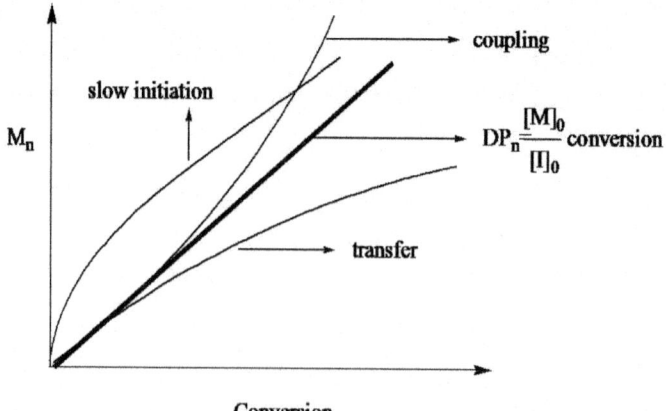

Conversion

FEATURE 3 Designed, usually narrow, molecular weight distribution:

It is important to recognize that the evolution of molecular weight is not very sensitive to chain termination, since the number of chains remains unchanged. The effect of termination is only observable on the plot when coupling reactions, forming higher molecular weight polymers, start to play a significant role.

Although this feature is often desirable, it is not necessarily the result of a controlled polymerization, which only requires the absence of chain transfer and termination, but ignores the effect of rate of initiation, exchange and depropagation.

Previous studies indicate that in order to obtain a polymer with a narrow molecular weight distribution, each of the following requirements should be fulfilled.

i.   *The rate of initiation is fast in comparison to the rate of propagation.* This condition allows simultaneous growth of all the polymer chains.

ii. *The exchange between species of different reactivity is fast in comparison with the rate of propagation.* This condition ensures that all the active chain termini are equally susceptible to reaction with monomer allowing uniform chain growth.

iii. *There must be negligible chain transfer or termination.*

iv. *The rate of depropagation is substantially lower than propagation.* This guarantees that polymerization is essentially irreversible.

This should yield a Poisson distribution, as quantified in equation CRP.4.

$$\frac{X_w}{X_n} = \frac{M_w}{M_n} = 1 + \frac{X_n}{\left(X_n + 1\right)^2} \cong + \frac{1}{X_n} \qquad \text{(CRP.4)}$$

According to equation CRP.4, polydispersity (PDI = M_w / M_n) decreases with increasing molecular weight.

Systems with slow exchange do not follow this perfect distribution but PDI's in a well conducted ATRP are defined by the following equation, where k_p is the rate of propagation, k_{deact} the rate of deactivation, [RX] the concentration of initiator and dormant chain ends and [Cu^{II}L$_n$X] is the concentration of the higher oxidation state transition metal complex, the deactivator.

$$\text{PDI} = \frac{M_w}{M_n} = 1 + \left(\frac{k_p [RX]_0}{K_{deact}\left[Cu^{II}L_nX\right]}\right)\left(\frac{2}{conv} - 1\right)$$

A polymerization that satisfies the prerequisites listed above is expected to form a final polymer with a polydispersity less than 1.1 for X_n greater than 10. However, it is possible to fine tune PDI by varying [Cu^{II}L$_n$X] or conversion. Well-controlled segmented copolymers with higher PDI can self-organize to materials with interesting nanostructured morphologies.

This is a consequence of negligible irreversible chain transfer and termination. Hence, all the chains retain their active centers after the full consumption of the monomer. Propagation resumes upon introduction of additional monomer.

This unique feature enables the preparation of block copolymers by sequential monomer addition.

The significance of controlled polymerization as a synthetic tool is widely recognized and polymers having uniform predictable chain length are readily available. Controlled polymerization provides the best opportunity to control the bulk properties of a target material through control of the multitude of possible variations in composition, functionality and topology now attainable at a molecular level.

Through appropriate selection of the functional (macro)initiator, copolymers formed in a controlled/"living" polymerization process can have essentially any desired topology. Further, as noted at the foot of the figure showing what CRP can do, we highlight that mechanistic transformations permit the use of macroinitiators or macromonomers prepared by other polymerization procedures in many CRP processes which allows incorporation of a spectrum of functionalities and polymer segments prepared by any other controlled polymerization process into segments of copolymers prepared by CRP.

PROCESS OPTIMIZATION

Process optimization is the discipline of adjusting a process so as to optimize some specified set of parameters without violating some constraint. The most common goals are minimizing cost, maximizing throughput, and/or efficiency. This is one of the major quantitative tools in industrial decision making.

When optimizing a process, the goal is to maximize one or more of the process specifications, while keeping all others within their constraints.

Areas

Fundamentally, there are three parameters that can be adjusted to affect optimal performance. They are:

- Equipment optimization: The first step is to verify that the existing equipment is being used to its fullest advantage by examining operating data to identify equipment bottlenecks.

- Operating procedures: Operating procedures may vary widely from person-to-person or from shift-to-shift. Automation of the plant can help significantly. But automation will be of no help if the operators take control and run the plant in manual.

- Control optimization: In a typical processing plant, such as a chemical plant or oil refinery, there are hundreds or even thousands of control loops. Each control loop is responsible for controlling one part of the process, such as maintaining a temperature, level, or flow.

If the control loop is not properly designed and tuned, the process runs below its optimum. The process will be more expensive to operate, and equipment will wear out prematurely. For each control loop to run optimally, identification of sensor, valve, and tuning problems is important. It has been well documented that over 35% of control loops typically have problems.

The process of continuously monitoring and optimizing the entire plant is sometimes called performance supervision.

MODEL BUILDING AND OPTIMIZATION: GENERAL FRAMEWORK AND WORKFLOW

Buildingpoint for General Contractors and Construction Managers

General contractors are under increased pressure to deliver in the "build" phase of today's complex design-build-operate (DBO) lifecycle on major construction projects. BuildingPoint gives you the advantage with a portfolio of building-intelligence software, hardware, and services solutions that optimizes workflows, streamlines processes, and seamlessly fits within existing operational frameworks. Our solutions operate within your business structures and processes, leveraging technology that won't distract you from your core competency of building.

Our Virtual Design and Construction (VDC) services assist general contractors during times of peak demand during a project. We can optimize any package to include business development support; 2Dchange reports (between two document sets); 3D model, constructability and coordination; 4D schedule simulation and optimization, 5D model-based quantity takeoff and estimates; laser scanning and as built models.

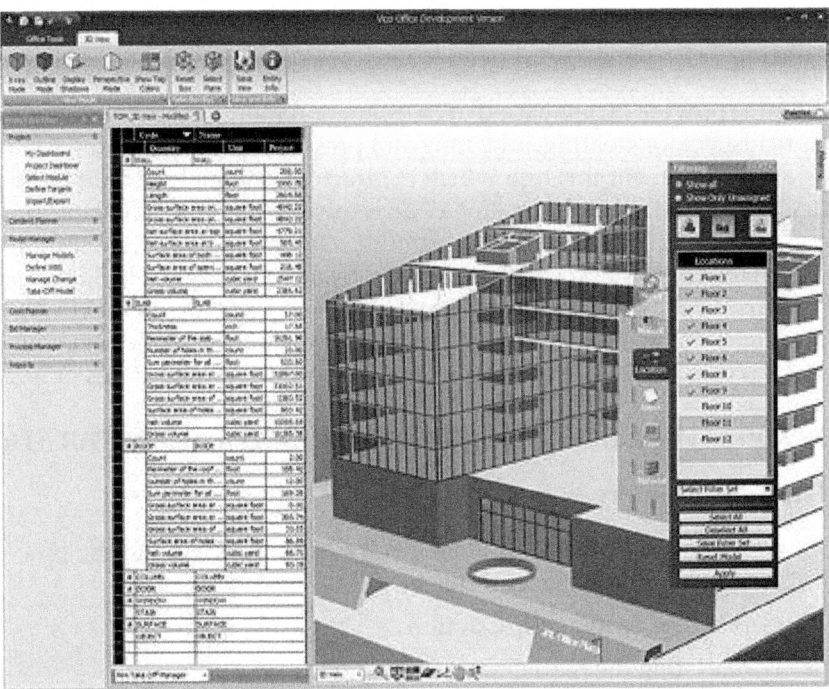

Virtual Construction

BuildingPoint's Vico Office offers a different way of working with Building Information Modeling (BIM) models. Used for much more than simple visualization, Vico Office extends the basic 3D model with constructability analysis and coordination, quantity takeoff, 4D location-based scheduling and production control, and 5D estimating.

Project Workflow

BuildingPoint's Prolog offers cost control, streamlining the project information workflow and providing access to information from anywhere. Prolog gives the operator the flexibility to access construction project data through a variety of interfaces and devices, so that your project team gets critical information when and where they need it. Desktop, Web and Mobile solutions give contractors and their teams the option to work online or offline; in the office or out in the field.

Natively built for mobile devices, Prolog Mobile allows you to access and capture critical Prolog project information straight from your iPad or Windows Mobile device. Prolog Mobile empowers field staff by extending the power of Prolog onto mobile platforms without requiring special integration. Field staff can view current project information and even manage construction oversight activities, with or without an Internet connection. Data is collected in real time from the field and seamlessly synchronized with Prolog back in the office, creating a complete system of record for all project data.

Estimating

WinEst provides a database-driven estimation solution that uses a highly-flexible spreadsheet for creating, adjusting and presenting cost estimates. Modelogix offers an environment through which contractors can analyze their project cost history for fast and accurate creation of early-phase budgeting and benchmarking of final estimates. Many features streamline estimation and make it more accurate. The New Metrics provides more than twenty customizable data types and cost centers — allowing you to intelligently create your cost model in ways that are most meaningful to your organization. Inflationary cost indexes allow you to normalize costs in all your past projects — regardless of their dates or locations.

OPTIMIZATION: DEFINITION AND MATHEMATICAL FORMULATION

The Optimisation Process

Solving an optimisation problem is not a linear process, but the process can be broken down into five general steps:
- Getting the problem description
- Formulating the mathematical program

- Solving the mathematical program
- Performing some post-optimal analysis
- Presenting the solution and analysis

However, there are often "feedback loops" within this process. For example, after formulating and solving an optimisation problem, you will often want to consider the validity of your solution (often consulting with the person who provided the problem description). If your solution is invalid you may need to alter or update your formulation to incorporate your new understanding of the actual problem. This process is shown in the Operations Research Methodology Diagram.

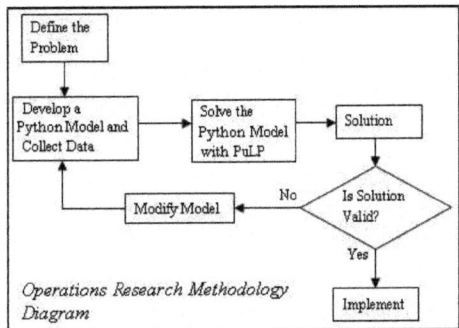

Operations Research Methodology Diagram

The modeling process starts with a well-defined model description, then uses mathematics to formulate a mathematical program. Next, the modeler enters the mathematical program into some solver software, *e.g.*, Excel and solves the model. Finally, the solution is translated into a decision in terms of the original model description.

Using Python gives you a "shortcut" through the modeling process. By formulating the mathematical program in Python you have already put it into a form that can be used easily by PuLP the modeller to call many solvers, *e.g.* CPLEX, COIN, gurobi so you don't need to enter the mathematical program into the solver software. However, you usually don't put any "hard" numbers into your formulation, instead you "populate" your model using data files, so there is some work involved in creating the appropriate data file. The advantage of using data files is that the same model may used many times with different data sets.

The Modeling Process

The modeling process is a "neat and tidy" simplification of the optimisation process.

Getting the Problem Description

The aim of this step is to come up with a formal, rigourous model description. Usually you start an optimisation project with an abstract description of a problem and some data. Often you need to spend some time talking with the person providing the problem (usually known as the client). By talking with the client and considering the data available you can come up with the more rigourous model description you are used to. Sometimes not all the data will be relevant or you will need to ask the client if they can provide some other data. Sometimes the limitations of the available data may change your model description and subsequent formulation significantly.

Formulating the mathematical program

In this step we identify the key quantifiable decisions, restrictions and goals from the problem description, and capture their interdependencies in a mathematical model. We can break the formulation process into 4 key steps:

- Identify the Decision Variables paying particular attention to units (for example: we need to decide how many hours per week each process will run for).
- Formulate the Objective Function using the decision variables, we can construct a minimise or maximise objective function. The objective function typically reflects the total cost, or total profit, for a given value of the decision variables.
- Formulate the Constraints, either logical (for example, we cannot work for a negative number of hours), or explicit to the problem description. Again, the constraints are expressed in terms of the decision variables.
- Identify the Data needed for the objective function and constraints. To solve your mathematical program you will need to have some "hard numbers" as variable bounds and/or variable coefficients in your objective function and/or constraints.

Solving the Mathematical Program

For relatively simple or well understood problems the mathematical model can often be solved to optimality (*i.e.*, the best possible solution is identified). This is done using algorithms such as the Revised Simplex Method or Interior Point Methods. However, many industrial problems would take too long to solve to optimality using these techniques, and so are solved using heuristic methods which do not guarantee optimality.

Performing Some Post-optimal Analysis

Often there is uncertainty in the problem description (either with the accuracy of the data provided, or with the value(s) of data in the future). In this situation the robustness of our solution can be examined by performing post-optimal analysis. This involves identifying how the optimal solution would change under various changes to the formulation (for example, what would be the effect of a given cost increasing, or a particular machine failing?). This sort of analysis can also be useful for making tactical or strategic decisions (for example, if we invested in opening another factory, what effect would this have on our revenue?).

Another important consideration in this step (and the next) is the validation of the mathematical program's solution. You should carefully consider what the solution's variable values mean in terms of the original problem description. Make sure they make sense to you and, more importantly, your client (which is why the next step, presenting the solution and analysis is important).

Presenting the Solution and Analysis

A crucial step in the optimisation process is the presentation of the solution and any post-optimal analysis. The translation from a mathematical program's solution back into a concise and comprehensible summary is as important as the translation from the problem description into the mathematical program. Key observations and decisions generated via optimisation must be presented in an easily understandable way for the client or project stakeholders.

Your presentation is a crucial first step in the implementation of the decisions generated by your mathematical program. If the decisions and their consequences (often determined by the mathematical program constraints) are not presented clearly and intelligently your optimal decision will never be used.

This step is also your chance to suggest other work in the future. This could include:

- Periodic monitoring of the validity of your mathematical program;
- Further analysis of your solution, looking for other benefits for your client;
- Identification of future optimisation opportunities.

Linear Programing

The simplest type of mathematical program is a linear program. For your mathematical program to be a linear program you need the following conditions to be true:

- The decision variables must be real variables;
- The objective must be a linear expression;
- The constraints must be linear expressions.

Linear expressions are any expression of the form

$$a_1 x_1 + a_2 x_2 + a_3 x_3 + \cdots a_n x_n \left\{ <=, =, >= \right\} b$$

where the a_i and b are known constants and x_i are variables. The process of solving a linear program is called linear programing. Linear programing is done via the Revised Simplex Method (also known as the Primal Simplex Method), the Dual Simplex Method or an Interior Point Method.

Integer Programing

Integer programs are almost identical to linear programs with one very important exception. Some of the decision variables in integer programs may need to have only integer values. The variables are known as integer variables. Since most integer programs contain a mix of continuous variables and integer variables they are often known as mixed integer programs. While the change from linear programing is a minor one, the effect on the solution process is enormous. Integer programs can be very difficult problems to solve and there is a lot of current research finding "good" ways to solve integer programs. Integer programs can be solved using the branch-and-bound process.

FORMULATION OF OPTIMIZATION PROBLEMS

Three basic components are required to optimize an industrial process. First, the process or a mathematical model of the process must be available, and the process variables which can be manipulated and controlled must be known. Often, obtaining a satisfactory process model with known control variables is the most difficult task. Secondly, an economic model of the process is required. This is an equation that represents the profit made from the sale of products and costs associated with their production, such as raw materials, operating costs, fixed costs, taxes, *etc.* Finally, an optimization procedure must be selected which locates the values of the independent variables of the process to produce the maximum profit or minimum cost as measured by the economic model. Also, the constraints in materials, process equipment, manpower, *etc.* must be satisfied as specified in the process model.

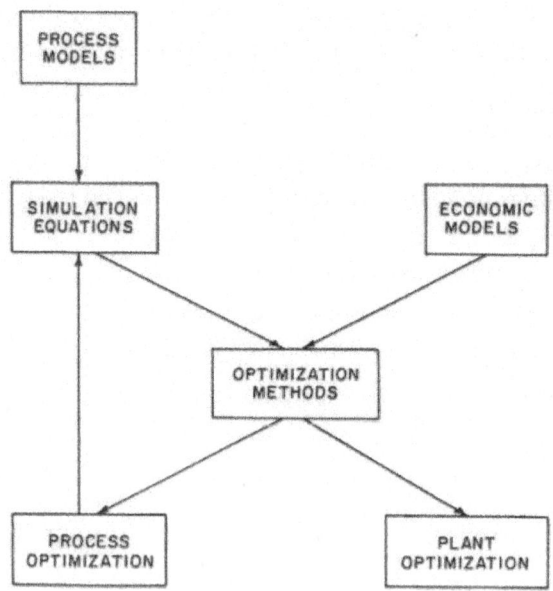

Fig. : Simplified Diagram of Industrial practice for precess and Plant Optimization.

The place industrial practice in perspective by relating process and economic models and the two levels of optimization. Plant optimization finds the best operating conditions for a plant made up of process units manufacturing specified amounts of various products to maximize the company's profits within the constraints set by the available raw materials and how these raw materials can be transformed in the plant. Plant optimization usually approximates the individual process units in a relatively simple manner to obtain a satisfactory answer in a reasonable time. This requires that the optimal operating conditions of the individual process unit be known, and these results be used in the plant optimization to have the plant operating with the maximum profit. Also, due to the complexity of large industrial plants, individual process models are usually simplified by using simulation equations to keep the computer programming efforts and computer costs within reason. However, with individual process units it is feasible to use more detailed models to determine more precisely the optimal operating conditions, *e.g.*, temperatures, pressures, recycle rates, *etc.* to have minimum operating cost known as a function of these variables.

The simulation equations are obtained from process models. The procedure is to develop precise process models based on the fundamentals of thermodynamics, kinetics and transport phenomena. This usually leads to process models which accurately represent the physical and chemical changes taking place over a wide range of conditions. However, these models usually are more complicated in mathematical form and may require the solution of differential equations. Consequently, these process models are usually exercised over the range of operation of the process, and simulation (regression) equations of a simplified mathematical form are developed, which are then used with the optimization method for the plant optimization. However, it may not be necessary to go through the simulation equations step if the equation that describe the key variables, *i.e.*, the ones that effect the economic performance of the process or plant, are not complicated.

Topics in Optimization

The two areas of optimization theory are mathematical programming and variational methods. Also, a number of techniques are listed under each of these areas. In mathematical programming, the objective is to locate a best point x $(x_1, x_2, ...x_n)$ which optimizes (maximizes or minimizes) the economic model of the process. In variational methods, the objective is to locate the optimal function which maximizes or minimizes the economic model. An example of an optimization problem for each division is given in the figure. Generally, mathematical programming methods are applicable to steady-state problems, and variational methods are for dynamic problems.

Mathematical programming methods are of two types and are referred to as direct or indirect methods. Direct methods, such as multivariable search methods and linear programming, move from a starting point through consistently improved values of the economic model to arrive at the optimum. Indirect methods, such as analytical methods and geometric programming, solve a set of algebraic equations, and the solution to the set of equations may be the optimum of the economic model. For example, in analytical methods the algebraic equation set is obtained by differentiating the economic model with respect to each independent variable and setting the resulting equations equal to zero.

Analytical methods are also called the classical theory of maxima and minima which is concerned with finding the extreme points of a function. The unconstrained and constrained optimization problems. Geometric programming may be considered an extension of analytical methods where the economic model and constraints are polynomials, and a dual problem is constructed that may be significantly easier to optimize than the original or primal problem.

Optimization

Areas and Topics in Optimization.

Mathematical Programming	Variational Methods
Objective: Fine the best point that optimizes the economic model. Example: Optimum operating conditions for a petroleum refinery.	Objective: Find the best function that optimizes the economic model. Example: Best temperature profile that maximizes the conversion in a tubular chemical reactor.
Mathematical Formula:	**Mathematical Formulation:**
Optimize: $y(x)$ Subject to: $f\,i(x) > 0$ $i = 1,2,...m$ where $x = (x_1, x_2...x_n)$	Optimize: $I\,[y(x)] = IF[y(x), y'(x)]dx$ Subject to: Algebraic, integral or differential equation constraints.
Methods	**Methods**
Analytical Methods Geometric Programming Linear Programming Quadratic Programming Convex Programming Dynamic Programming (Discrete) Nonlinear Programming or Multivariable Search Methods Integer Programming Separable Programming Goal Programming or Multicriterion Optimization Combinatorial Programming Maximum Principle (Discrete) Heuristic Programming	Calculus of Variations Dynamic Programming(Continuous) Maximum Principle (Continuous)

Linear programming requires that both the economic model and the set of constraint equations be linear, and the Simplex Method is the algorithm which locates the optimum by beginning at a feasible starting point (initially feasible basis). In quadratic programming, the economic model is a quadratic equation and the constraint equations are linear. Using analytical methods, this problem can be converted to a linear programming problem and solved by the Simplex Method. For convex programming, the economic model is a concave function, and the constraint equations are convex functions. The details on this procedure of general analytical methods and show that a global optimum will be located.

Dynamic programming uses a series of partial optimizations by taking advantage of the stage structure in the problem and is effective for resource allocation and optimization through time. Nonlinear programming or multivariable search methods, as the theory and algorithms are called, must begin at a feasible starting point and move toward the optimum in steps of improved values of the economic model. The algorithms have been effective for optimization of industrial processes, and they are based on the theory.

Integer programming is an extension of linear programming where the variables must take on discrete values, and a text on this topic is by Taha. Separable programming is an extension of linear programming where a small number of nonlinear constraints are approximated by piecewise linear functions. However, the nonlinear functions must have the form so they can be separated into sums and differences of nonlinear functions of one variable, and the IBM MPSX code is capable of solving these problems. Goal programming is an extension of linear programming also where multiple, conflicting objectives (or goals) are optimized using weights or rankings, for example, and this technique is described by Ignizio. Combinatorial programming has been described by Popadimitriou and Steiglitz as a body of mathematical programming knowledge including linear and integer programming, graph and network flows, dynamic programming and related topics. The maximum principle is comparable to dynamic programming in using the stage structure of the system, but it uses constrained derivatives that require piecewise continuously differentiable functions and successive approximations. Finally, the term heuristic programming has been used to describe rules-of-thumb that can be used for approximations to optimization.

The various topics in optimization, the economic model has been given several different names. These names arose in the literature as the optimization procedures were being developed. Regardless of the name, the economic model is the equation which expresses the economic return from the process for specified values of the control (manipulative, decision or independent) variables. The two most common names are the profit function or cost function. However, in linear programming the term objective function is used, and in dynamic programming the term return function is employed. Other synonymous names are: benefit function, criterion, measure of effectiveness and response surface.

Method of Attack

In solving an optimization problem, the structure and complexity of the equations for the economic model and process or plant constraints are very important, since most mathematical programming procedures take advantage of the mathematical form of these models. Examples are linear programming, where all of the equations must be linear, and geometric programming, where all of the equations must be polynomials. Consequently, it is extremely important to have the capabilities of the various optimization techniques in mind when the economic and process models are being formulated. For example, if a satisfactory representation of the economics and process performance can be obtained using all linear equations, the powerful techniques of linear programming can be applied, and this method guarantees that a global optimum is found. However, if one has to resort to nonlinear equations to represent the economics and process performance, it may be necessary to use a multivariable search method to locate the optimum. Unfortunately, these search techniques only find points that are better than the starting point, and they do not carry any guarantee that a global or a local maximum or minimum has been found.

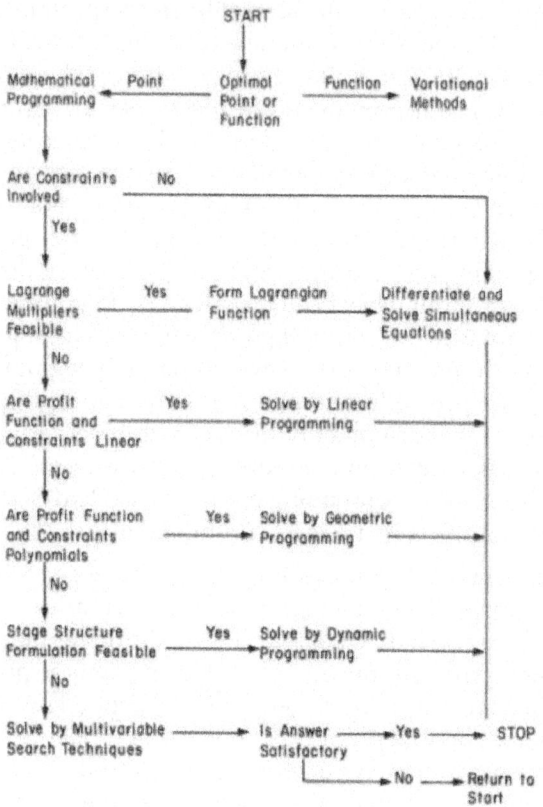

Fig. : Simplified Method of Attach for Optimization Problems.

A simplified approach to attacking an optimization problem, and it incorporates some thoughts which should be remembered as the particular optimization techniques are studied. Also, it will give some reasons for the order in which the techniques are presented. At the start, it is necessary to determine if the problem requires an optimal point or function. If it is a point, mathematical programming is applicable; and if an optimal function, variational methods. Let us follow through with mathematical programming. If the equation for the economic model is relatively simple and there are no applicable constraints (process model), it is possible to locate the optimum by differentiating the economic model with respect to the independent variables, setting these equations equal to zero, and solving for the optimum. However, if there are constraints, and there usually are, but the equations are relatively simple, the method of Lagrange multipliers may be used. This converts the constrained problem to an unconstrained problem, and the previous procedure for unconstrained problems is used.

Now, if the problem has a large number of independent variables and the precision needed for the economic and process models can be obtained with linear equations, then linear programming may be used. However, if nonlinear equations are required and polynomial will suffice, it may be possible to determine the optimum rapidly and easily using geometric programming.

Not having been successful to this point, it may be feasible to take advantage of the stage structure of the problem and apply dynamic programming with a series of partial optimizations. However, if this is not successful it will be necessary to resort to multivariable search techniques and seek best values without having a guarantee of finding the global optimum.

MAIN CLASSES OF OPTIMIZATION PROBLEMS

Types of Optimization Problems

As noted in the Introduction to Optimization, an important step in the optimization process is classifying your optimization model, since algorithms for solving optimization problems are tailored to a particular type of problem. Here we provide some guidance to help you classify your optimization model; for the various optimization problem types, we provide a linked page with some basic information, links to algorithms and software, and online and print resources.

- *Continuous Optimization* versus *Discrete Optimization*
- Some models only make sense if the variables take on values from a discrete set, often a subset of integers, whereas other models contain variables that can take on any real value. Models with discrete variables are *discrete optimization* problems; models with continuous variables are *continuous optimization* problems. Continuous optimization problems tend to be easier to solve than discrete optimization problems; the smoothness of the functions means that the objective function and constraint function values at a point x can be used to deduce information about points in a neighborhood of x. However, improvements in algorithms coupled with advancements in computing technology have dramatically increased the size and complexity of discrete optimization problems that can be solved efficiently. Continuous optimization algorithms are important in discrete optimization because many discrete optimization algorithms generate a sequence of continuous subproblems.
- *Unconstrained Optimization* versus *Constrained Optimization*
- Another important distinction is between problems in which there are *no constraints* on the variables and problems in which there are *constraints* on the variables. *Unconstrained optimization* problems arise directly in

many practical applications; they also arise in the reformulation of *constrained* optimization problems in which the constraints are replaced by a penalty term in the objective function. *Constrained optimization* problems arise from applications in which there are explicit constraints on the variables. The constraints on the variables can vary widely from simple bounds to systems of equalities and inequalities that model complex relationships among the variables. Constrained optimization problems can be furthered classified according to the nature of the constraints (*e.g.,* linear, nonlinear, convex) and the smoothness of the functions (*e.g.,* differentiable or nondifferentiable).

- *None, One or Many Objectives*
- Most optimization problems have a single objective function. There are interesting cases when optimization problems have no objective function or multiple objective functions. *Feasibility problems* are problems in which the goal is to find values for the variables that satisfy the constraints of a model with no particular objective to optimize. *Complementarity problems* are pervasive in engineering and economics. The goal is to find a solution that satisfies the complementarity conditions. *Multi-objective optimization* problems arise in many fields, such as engineering, economics, and logistics, when optimal decisions need to be taken in the presence of trade-offs between two or more conflicting objectives. For example, developing a new component might involve minimizing weight while maximizing strength or choosing a portfolio might involve maximizing the expected return while minimizing the risk. In practice, problems with multiple objectives often are reformulated as single objective problems by either forming a weighted combination of the different objectives or by replacing some of the objectives by constraints.

- *Deterministic Optimization* versus *Stochastic Optimization*
- In *deterministic optimization*, it is assumed that the data for the given problem are known accurately. However, for many actual problems, the data cannot be known accurately for a variety of reasons. The first reason is due to simple measurement error. The second and more fundamental reason is that some data represent information about the future (e. g., product demand or price for a future time period) and simply cannot be known with certainty. In *optimization under uncertainty*, or *stochastic optimization*, the uncertainty is incorporated into the model. *Robust optimization* techniques can be used when the parameters are known only within certain bounds; the goal is to find a solution that is feasible for all data and optimal in some sense.Stochastic programming models take advantage of the fact that probability distributions governing the data are known or can be estimated; the goal is to find some policy that is feasible for all (or almost all) the possible data instances and optimizes the expected performance of the model.

Some Classes of Optimization Problems

Least-squares

The least-squares problem is

$$\min_x \sum_{i=1}^{m} \left(\sum_{j=1}^{n} A_{ij} x_j - b_i \right)^2 .$$

where $A_{ij}, b_i < 1 \leq i \leq m, 1 \leq j \leq n$, are given numbers, and $x \in R^n$ is the variable.

The problem was posed and solved by Gauss (1777-1855), who used the method it to predict the trajectory of the planetoid Ceres. Least-squares problems arise in many situations, for example in statistical estimation problems such as linear regression. In image compression this problem also arises, in which case b_i's contain a pixel representation of a given image, and for every j, the numbers A_{ij}, $i = 1,..., m$ contain representations of "basic" images; the sums inside the squares represents a linear combination of these basic images that is supposed to approximate as well as possible the original image contained in b_i's.

Linear Programming

This problem is often referred to with the acronym LP, and has the form

$$\min \sum_{j=1}^{m} c_j x_j \ : \ \sum_{j=1}^{n} A_{ij} x_j \leq b_i, i = 1, \ldots, m,$$

where c_j, b_i and $A_{ij}, 1 \leq i \leq m, 1 \leq j \leq n$, are given real numbers. This corresponds to the case where the functions $f_i(i = 0,..., m)$ in the standard problem are all affine (that is, linear plus a constant term).

This problem was introduced by G. Dantzig in the 40's in the context of logistical problems arising in military operations. This model of computation is perhaps the most widely used optimization problem today.

Quadratic Programming

Quadratic programming problems (QP's for short) are an extension of linear programming, which involve a sum-of-squares function in the objective. The linear program above is modified to be

$$\min \sum_{j=1}^{n} (x_i^2 + c_j x_j) \ : \ \sum_{j=1}^{n} A_{ij} x_j \leq b_i, \ i = 1, \ldots, m,$$

These problems can be thought of as a generalization of both the least-squares and linear programming problems.

QP's are popular in many areas, such as finance, where the linear term in the objective refers to the expected negative return on an investment, and the squared term corresponds to the risk (or variance of the return). This model was introduced in the 50's by H. Markowitz (who was then a colleague of Dantzig at the RAND Corporation), to model investment problems. Markowitz won the Nobel prize in Economics in 1990, mainly for this work.

Nonlinear Optimization

Nonlinear optimization is perhaps the largest class of optimization problem. In general such problems are very hard to solve. (In fact, this class comprises combinatorial optimization: if a variable x_i is required to be boolean, we can model this as a single non-linear constraint $x_i^2 = x_i$.)

One of the reasons for which non-linear problems are hard to solve is the issue of *local minima*. We will define this notion rigorously, but the picture below provides an intuitive idea:

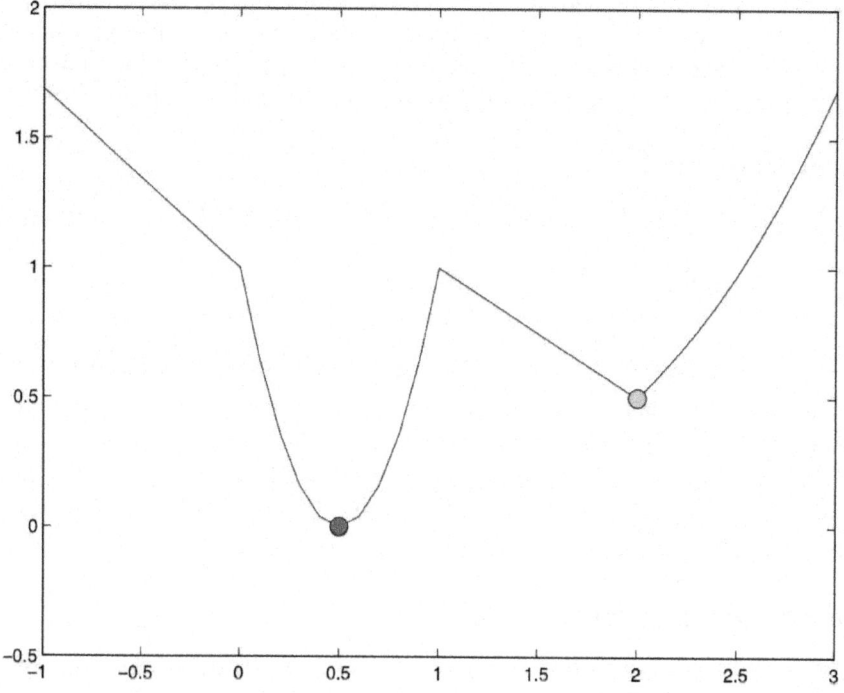

When trying to minimize general non-linear functions, algorithms may be trapped in so-called "local" minima, which do not correspond to the true minimal value of the objective function.

Convex Optimization

Convex optimization is a generalization of QP where the objective and constraints involve "bowl-shaped", or convex, functions. Not all convex problems are easy to solve, but many of them are. One of the reasons for which this is (approximately) true is that, contrarily to general non-linear optimization problems, convex ones do not suffer from the "curse of local minima" mentioned above.

When trying to minimize convex (that is, bowl-shaped) functions, specialized algorithms will always converge to a global minimum, irrespective of the starting point, provided some (weak) assumptions on the function hold.

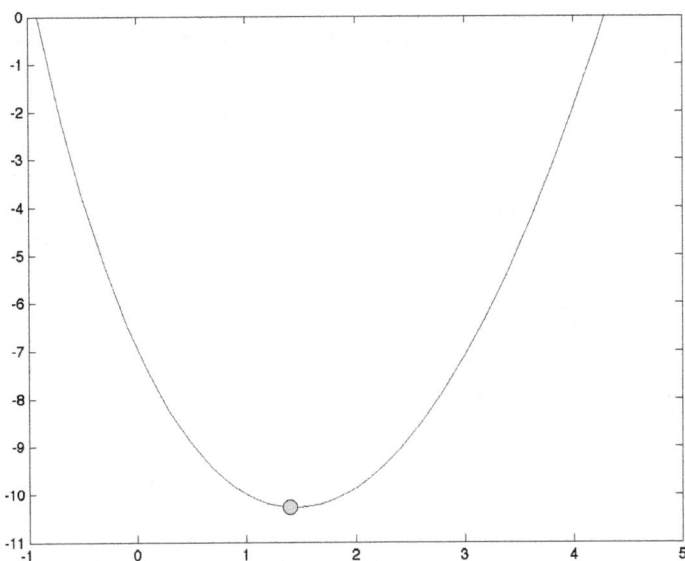

Combinatorial Optimization

In combinatorial optimization, some (or all) the variables are boolean (or integers), reflecting discrete choices to be made.

Combinatorial optimization problems are in general extremely hard to solve. Often, they can be approximately solved with linear or convex programming.

Optimization Problem

In mathematics and computer science, an optimization problem is the problem of finding the *best* solution from all feasible solutions. Optimization problems can be divided into two categories depending on whether the variables are continuous or discrete. An optimization problem with discrete variables is known as a combinatorial optimization problem. In a combinatorial optimization problem, we are looking for an object such as an integer, permutation or graph from a finite (or possibly countable infinite) set. Problems with continuous variables include constrained problems and multimodal problems.

Continuous Optimization Problem

The *standard form* of a (continuous) optimization problem is

$$\underset{x}{\text{minimize}} \quad f(x)$$

$$\text{subject to} \quad g_i(x) \le 0, \quad i = 1, \ldots, m$$
$$h_i(x) = 0, \quad i = 1, \ldots, p$$

where

- $f(x): \mathbb{R}^n \to \mathbb{R}$ is the objective function to be minimized over the variable x,

- $g_i(x) \le 0$ are called inequality constraints, and

- $h_i(x) = 0$ are called equality constraints.

By convention, the standard form defines a minimization problem. A maximization problem can be treated by negating the objective function.

Combinatorial Optimization Problem

Formally, a combinatorial optimization problem A is a quadruple (I, f, m, g), where

- I is a set of instances;
- given an instance $x \in I$, $f(x)$ is the set of feasible solutions;
- given an instance x and a feasible solution y of x, $m(x, y)$ denotes the measure of y, which is usually a positive real.
- g is the goal function, and is either min or max.

The goal is then to find for some instance x an *optimal solution*, that is, a feasible solution y with

$$m(x, y) = g\{m(x, y') \mid y' \in f(x)\}.$$

For each combinatorial optimization problem, there is a corresponding decision problem that asks whether there is a feasible solution for some particular measure m_0. For example, if there is a graph G which contains vertices u and v, an optimization problem might be "find a path from u to v that uses the fewest edges". This problem might have an answer of, say, 4. A corresponding decision problem would be "is there a path from u to v that uses 10 or fewer edges?" This problem can be answered with a simple 'yes' or 'no'.

In the field of approximation algorithms, algorithms are designed to find near-optimal solutions to hard problems. The usual decision version is then an inadequate definition of the problem since it only specifies acceptable solutions. Even though we could introduce suitable decision problems, the problem is more naturally characterized as an optimization problem.

NP Optimization Problem

An *NP-optimization problem* (NPO) is a combinatorial optimization problem with the following additional conditions. Note that the below referred polynomials are functions of the size of the respective functions' inputs, not the size of some implicit set of input instances.

- the size of every feasible solution $y \in f(x)$ is polynomially bounded in the size of the given instance x,
- the languages $\{x \mid x \in I\}$ and $\{(x, y) \mid y \in f(x)\}$ can be recognized in polynomial time, and
- m is polynomial-time computable.

This implies that the corresponding decision problem is in NP. In computer science, interesting optimization problems usually have the above properties and are therefore NPO problems. A problem is additionally called a P-optimization (PO) problem, if there exists an algorithm which finds optimal solutions in polynomial time. Often, when dealing with the class NPO, one is interested in optimization problems for which the decision versions are NP-complete. Note that hardness relations are always with respect to some reduction. Due to the connection between approximation algorithms and computational optimization problems, reductions which preserve approximation in some respect are for this subject preferred than the usual Turing and Karp reductions. An example of such a reduction would be the L-reduction. For this reason, optimization problems with NP-complete decision versions are not necessarily called NPO-complete.

NPO is divided into the following subclasses according to their approxima-bility:

- *NPO(I)*: Equals FPTAS. Contains the Knapsack problem.
- *NPO(II)*: Equals PTAS. Contains the Makespan scheduling problem.
- *NPO(III)*: :The class of NPO problems that have polynomial-time algo-rithms which computes solutions with a cost at most c times the optimal cost (for minimization problems) or a cost at least $1/c$ of the optimal cost (for maximization problems). In Hromkovič's book, excluded from this class are all NPO(II)-problems save if P=NP. Without the exclusion, equals APX. Contains MAX-SAT and metric TSP.
- *NPO(IV)*: :The class of NPO problems with polynomial-time algorithms approximating the optimal solution by a ratio that is polynomial in a loga-rithm of the size of the input. In Hromkovic's book, all NPO(III)-problems are excluded from this class unless P=NP. Contains the set cover problem.
- *NPO(V)*: :The class of NPO problems with polynomial-time algorithms approximating the optimal solution by a ratio bounded by some function on n. In Hromkovic's book, all NPO(IV)-problems are excluded from this class unless P=NP. Contains the TSP and Max Clique problems.

Another class of interest is NPOPB, NPO with polynomially bounded cost functions. Problems with this condition have many desirable properties.

INGREDIENTS OF OPTIMIZATION PROBLEMS AND
THEIR CLASSIFICATION

The essence of all businesslike decisions, whether made for a firm, or an individual, is finding a course of action that leaves you with the largest profit.

Mankind has long sought, or professed to seek, better ways to carry out the daily tasks of life. Throughout human history, man has first searched for more effective sources of food and then later searched for materials, power, and mastery of the physical environment. However, relatively late in human history general questions began to quantitatively formulate first in words, and later developing into symbolic notations. One pervasive aspect of these general questions was to seek the "best" or "optimum". Most of the time managers seek merely to obtain some improvement in the level of performance, or a "goal-seeking" problem. It should be emphasized that these words do not usually have precise meanings.

Efforts have been made to describe complex human and social situations. To have meaning, the problem should be written down in a mathematical expression containing one or more variables, in which the value of variables are to be determined. The question then asked, is what values should these variables have to ensure the mathematical expression has the greatest possible numerical value (maximization) or the least possible numerical value (minimization). This process of maximizing or minimizing is referred to as optimization.

Optimization, also called mathematical programming, helps find the answer that yields the best result--the one that attains the highest profit, output, or happiness, or the one that achieves the lowest cost, waste, or discomfort. Often these problems involve making the most efficient use of resources--including money, time, machinery, staff, inventory, and more. Optimization problems are often classified as linear or nonlinear, depending on whether the relationship in the problem is linear with respect to the variables. There are a variety of software packages to solve optimization problems. For example, LINDO or your WinQSB solve linear program models and LINGO and What's Best! solve nonlinear and linear problems.

Mathematical Programming, solves the problem of determining the optimal allocations of limited resources required to meet a given objective. The objective must represent the goal of the decision-maker. For example, the resources may correspond to people, materials, money, or land. Out of all permissible allocations of the resources, it is desired to find the one or ones that maximize or minimize some numerical quantity such as profit or cost. Optimization models are also called Prescriptive or Normative models since they seek to find the best possible strategy for decision-maker.

There are many optimization algorithms available. However, some methods are only appropriate for certain types of problems. It is important to be able to recognize the characteristics of a problem and identify an appropriate solution technique. Within each class of problems, there are different minimization methods, which vary in computational requirements, convergence properties, and so on.

Optimization problems are classified according to the mathematical characteristics of the objective function, the constraints, and the controllable decision variables.

Optimization problems are made up of three basic ingredients:

1. An objective function that we want to minimize or maximize. That is, the quantity you want to maximize or minimize is called the objective function. Most optimization problems have a single objective function, if they do not, they can often be reformulated so that they do. The two interesting exceptions to this rule are:

   The goal seeking problem: In most business applications the manager wishes to achieve a specific goal, while satisfying the constraints of the model. The user does not particularly want to optimize anything so there is no reason to define an objective function. This type of problem is usually called a feasibility problem.

   Multiple objective functions: Often, the user would actually like to optimize many different objectives at once. Usually, the different objectives are not compatible. The variables that optimize one objective may be far from optimal for the others. In practice, problems with multiple objectives are reformulated as single-objective problems by either forming a weighted combination of the different objectives or else by placing some objectives as "desirable" constraints.

2. The controllable inputs are the set of decision variables which affect the value of the objective function. In the manufacturing problem, the variables might include the allocation of different available resources, or the labor spent on each activity. Decision variables are essential. If there are no variables, we cannot define the objective function and the problem constraints.

3. The uncontrollable inputs are called parameters. The input values may be fixed numbers associated with the particular problem. We call these values parameters of the model. Often you will have several "cases" or variations of the same problem to solve, and the parameter values will change in each problem variation.

4. Constraints are relations between decision variables and the parameters. A set of constraints allows some of the decision variables to take on certain values, and exclude others. For the manufacturing problem, it does not make sense to spend a negative amount of time on any activity, so we constrain all the "time" variables to be non-negative. Constraints are not always essential. In fact, the field of unconstrained optimization is a large and important one for which a lot of algorithms and software are available. In practice, answers that make good sense about the underlying physical or economic problem, cannot often be obtained without putting constraints on the decision variables.

   Feasible and Optimal Solutions: A solution value for decision variables, where all of the constraints are satisfied, is called a feasible solution. Most solution algorithms proceed by first finding a feasible solution, then seeking to improve upon it, and finally changing the decision variables to move from one feasible solution to another feasible solution. This process is repeated until the objective function has reached its maximum or minimum. This result is called an optimal solution.

The basic goal of the optimization process is to find values of the variables that minimize or maximize the objective function while satisfying the constraints. This result is called an optimal solution.

There are well over 4000 solution algorithms for different kinds of optimization problems. The widely used solution algorithms are those developed for the following mathematical programs: convex programs, separable programs, quadratic programs and the geometric programs.

Linear Program

Linear programming deals with a class of optimization problems, where both the objective function to be optimized and all the constraints, are linear in terms of the decision variables.

A short history of Linear Programming:

1. In 1762, Lagrange solved tractable optimization problems with simple equality constraints.

2. In 1820, Gauss solved linear system of equations by what is now call Causssian elimination. In 1866 Wilhelm Jordan refinmened the method to finding least squared errors as ameasure of goodness-of-fit. Now it is referred to as the Gauss-Jordan Method.

3. In 1945, Digital computer emerged.

4. In 1947, Dantzig invented the Simplex Methods.

5. In 1968, Fiacco and McCormick introduced the Interior Point Method.

6. In 1984, Karmarkar applied the Interior Method to solve Linear Programs adding his innovative analysis.

Linear programming has proven to be an extremely powerful tool, both in modeling real-world problems and as a widely applicable mathematical theory. However, many interesting optimization problems are nonlinear. The study of such problems involves a diverse blend of linear algebra, multivariate calculus, numerical analysis, and computing techniques. Important areas include the design of computational algorithms (including interior point techniques for linear programming), the geometry and analysis of convex sets and functions, and the study of specially structured problems such as quadratic programming. Nonlinear optimization provides fundamental insights into mathematical analysis and is widely used in a variety of fields such as engineering design, regression analysis, inventory control, geophysical exploration, and economics.

Quadratic Program

Quadratic Program (QP) comprises an area of optimization whose broad range of applicability is second only to linear programs. A wide variety of applications fall naturally into the form of QP. The kinetic energy of a projectile is a quadratic function of its velocity. The least-square regression with side constraints has been modeled as a QP. Certain problems in production planning, location analysis, econometrics, activation analysis in chemical mixtures problem, and in financial portfolio management and selection are often treated as QP. There are numerous solution algorithms available for the case under the restricted additional condition, where the objective function is convex.

Constraint Satisfaction

Many industrial decision problems involving continuous constraints can be modeled as continuous constraint satisfaction and optimization problems. Constraint Satisfaction problems are large in size and in most cases involve transcendental functions. They are widely used in chemical processes and cost restrictions modeling and optimization.

Convex Program

Convex Program (CP) covers a broad class of optimization problems. When the objective function is convex and the feasible region is a convex set, both of these assumptions are enough to ensure that local minimum is a global minimum.

Data Envelopment Analysis

The Data Envelopment Analysis (DEA) is a performance metric that is grounded in the frontier analysis methods from the economics and finance literature. Frontier efficiency (output/input) analysis methods identify best practice performance frontier, which refers to the maximal outputs that can be obtained from a given set of inputs with respect to a sample of decision making units using a comparable process to transform inputs to outputs. The strength of DEA relies partly on the fact that it is a non-parametric approach, which does not require specification of any functional form of relationships between the inputs and the outputs. DEA output reduces multiple performance measures to a single one to use linear programming techniques. The weighting of performance measures reacts to the decision-maker's utility.

Dynamic Programming

Dynamic programming (DP) is essentially bottom-up recursion where you store the answers in a table starting from the base case(s) and building up to larger and larger parameters using the recursive rule(s). You would use this technique instead of recursion when you need to calculate the solutions to all the sub-problems and the recursive solution would solve some of the sub-problems repeatedly. While generally DP is capable of solving many diverse problems, it may require huge computer storage in most cases.

Separable Program

Separable Program (SP) includes a special case of convex programs, where the objective function and the constraints are separable functions, *i.e.*, each term involves just a single variable.

Geometric Program

Geometric Program (GP) belongs to Nonconvex programming, and has many applications in particular in engineering design problems.

Fractional Program

In this class of problems, the objective function is in the form of a fraction (*i.e.*, ratio of two functions). Fractional Program (FP) arises, for example, when maximizing the ratio of profit capital to capital expended, or as a performance measure wastage ratio.

Heuristic Optimization

A heuristic is something "providing aid in the direction of the solution of a problem but otherwise unjustified or incapable of justification." So heuristic arguments are used to show what we might later attempt to prove, or what we might expect to find in a computer run. They are, at best, educated guesses.

Several heuristic tools have evolved in the last decade that facilitate solving optimization problems that were previously difficult or impossible to solve. These tools include evolutionary computation, simulated annealing, tabu search, particle swarm, *etc.*

Common approaches include, but are not limited to:

1. comparing solution quality to optimum on benchmark problems with known optima, average difference from optimum, frequency with which the heuristic finds the optimum.

2. comparing solution quality to a best known bound for benchmark problems whose optimal solutions cannot be determined.

3. comparing your heuristic to published heuristics for the same problem type, difference in solution quality for a given run time and, if relevant, memory limit.

4. profiling average solution quality as a function of run time, for instance, plotting mean and either min and max or 5th and 95th percentiles of solution value as a function of time -- this assumes that one has multiple benchmark problem instances that are comparable.

Global Optimization

The aim of Global Optimization (GO) is to find the best solution of decision models, in presence of the multiple local solutions. While *constrained optimization* is dealing with finding the optimum of the objective function subject to constraints on its decision variables, in contrast, *unconstrained optimization* seeks the global maximum or minimum of a function over its entire domain space, without any restrictions on decision variables.

Nonconvex Program

A Nonconvex Program (NC) encompasses all nonlinear programming problems that do not satisfy the convexity assumptions. However, even if you are successful at finding a local minimum, there is no assurance that it will also be a global minimum. Therefore, there is no algorithm that will guarantee finding an optimal solution for all such problem.

Nonsmooth Program

Nonsmooth Programs (NSP) contain functions for which the first derivative does not exist. NSP are arising in several important applications of science and engineering, including contact phenomena in statics and dynamics or delamination effects in composites. These applications require the consideration of nonsmoothness and nonconvexity.

Metaheuristics

Most metaheuristics have been created for solving discrete combinatorial optimization problems. Practical applications in engineering, however, usually require techniques, which handle continuous variables, or miscellaneous continuous and discrete variables. As a consequence, a large research effort has focused on fitting several well-known metaheuristics, like Simulated Annealing (SA), Tabu Search (TS), Genetic Algorithms (GA), Ant Colony Optimization (ACO), to the continuous cases. The general metaheuristics aim at transforming discrete domains of application into continuous ones, by means of:

- Methodological developments aimed at adapting some metaheuristics, especially SA, TS, GA, ACO, GRASP, variable neighborhood search, guided local search, scatter search, to continuous or discrete/continuous variable problems.
- Theoretical and experimental studies on metaheuristics adapted to continuous optimization, *e.g.*, convergence analysis, performance evaluation methodology, test-case generators, constraint handling, *etc.*
- Software implementations and algorithms for metaheuristics adapted to continuous optimization.
- Real applications of discrete metaheuristics adapted to continuous optimization.
- Performance comparisons of discrete metaheuristics (adapted to continuous optimization) with that of competitive approaches, *e.g.*, Particle Swarm Optimization (PSO), Estimation of Distribution Algorithms (EDA),

Evolutionary Strategies (ES), specifically created for continuous optimization.

Multilevel Optimization

In many decision processes there is a hierarchy of decision makers and decisions are taken at different levels in thishierarchy. Multilevel Optimization focuses on the whole hierarchy structure. The field of multilevel optimization has become a well known and important research field. Hierarchical structures can be found in scientific disciplines such as environment, ecology, biology, chemical engineering, mechanics, classification theory, databases, network design, transportation, supply chain, game theory and economics. Moreover, new applications are constantly being introduced.

Multiobjective Program

Multiobjective Program (MP) known also as Goal Program, is where a single objective characteristic of an optimization problem is replaced by several goals. In solving MP, one may represent some of the goals as constraints to be satisfied, while the other objectives can be weighted to make a composite single objective function.

Multiple objective optimization differs from the single objective case in several ways:

1. The usual meaning of the optimum makes no sense in the multiple objective case because the solution optimizing all objectives simultaneously is, in general, impractical; instead, a search is launched for a feasible solution yielding the best compromise among objectives on a set of, so called, efficient solutions;

2. The identification of a best compromise solution requires taking into account the preferences expressed by the decision-maker;

3. The multiple objectives encountered in real-life problems are often mathematical functions of contrasting forms.

4. A key element of a goal programming model is the achievement function; that is, the function that measures the degree of minimisation of the unwanted deviation variables of the goals considered in the model.

A Business Application: In credit card portfolio management, predicting the cardholder's spending behavior is a key to reduce the risk of bankruptcy. Given a set of attributes for major aspects of credit cardholders and predefined classes for spending behaviors, one might construct a classification model by using multiple criteria linear programming to discover behavior patterns of credit cardholders.

Non-Binary Constraints Program

Over the years, the constraint programming community has paid considerable attention to modeling and solving problems by using binary constraints. Only recently has non-binary constraints captured attention, due to growing number of real-life applications. A non-binary constraint is a constraint that is defined on k variables, where k is normally greater than two. A non-binary constraint can be seen as a more global constraint. Modeling a problem as a non-binary constraint has two main advantages: It facilitates the expression of the problem; and it enables more powerful constraint propagation as more global information becomes available.

Success in timetabling, scheduling, and routing, has proven that the use of non-binary constraints is a promising direction of research. In fact, a growing number of OR/MS/DS workers feel that this topic is crucial to making constraint technology a realistic way to model and solve real-life problems.

Bilevel Optimization

Most of the mathematical programming models deal with decision-making with a single objective function. The bilevel programming on the other hand is developed for applications in decentralized planning systems in which the first level is termed as the leader and the second level pertains to the objective of the follower. In the bilevel programming problem, each decision maker tries to optimize its own objective function without considering the objective of the other party, but the decision of each party affects the objective value of the other party as well as the decision space.

Bilevel programming problems are hierarchical optimization problems where the constraints of one problem are defined in part by a second parametric optimization problem. If the second problem has a unique optimal solution for all parameter values, this problem is equivalent to usual optimization problem having an implicitly defined objective function. However, when the problem has non-unique optimal solutions, the optimistic (or weak) and the pessimistic (or strong) approaches are being applied.

Combinatorial Optimization

Combinatorial generally means that the state space is discrete (*e.g.*, symbols, not necessarily numbers). This space could be finite or denumerable sets. For example, a discrete problem is combinatorial. Problems where the state space is totally ordered can often be solved by mapping them to the integers and applying "numerical" methods. If the state space is unordered or only partially ordered, these methods fail. This means that the **heuristics methods** becomes necessary, such as simulated annealing.

Combinatorial optimization is the study of packing, covering, and partitioning, which are applications of integer programs. They are the principle mathematical topics in the interface between combinatorics and optimization. These problems deal with the classification of integer programming problems according to the complexity of known algorithms, and the design of good algorithms for solving special subclasses. In particular, problems of network flows, matching, and their matroid generalizations are studied. This subject is one of the unifying elements of combinatorics, optimization, operations research, and computer science.

Evolutionary Techniques

Nature is a robust optimizer. By analyzing nature's optimization mechanism we may find acceptable solution techniques to intractable problems. Two concepts that have most promise are simulated annealing and the genetic techniques. Scheduling and timetabling are amongst the most successful applications of evolutionary techniques.

Genetic Algorithms (GAs) have become a highly effective tool for solving hard optimization problems. However, its theoretical foundation is still rather fragmented.

Particle Swarm Optimization

Particle Swarm Optimization (PSO) is a stochastic, population-based optimization algorithm. Instead of competition/selection, like say in Evolutionary Computation, PSO makes use of cooperation, according to a paradigm sometimes called "swarm intelligence". Such systems are typically made up of a population of simple interacting agents without any centralized control, and inspired by cases that can be found in nature, such as ant colonies, bird flocking, animal herding, bacteria molding, fish schooling, *etc.*

There are many variants of PSO including constrained, multiobjective, and discrete or combinatorial versions, and applications have been developed using PSO in many fields.

Swarm Intelligence

Biologists studied the behavior of social insects for a long time. After millions of years of evolution all these species have developed incredible solutions for a wide range of problems. The intelligent solutions to problems naturally emerge from the self-organization and indirect communication of these individuals. Indirect interactions occur between two individuals when one of them modifies the environment and the other responds to the new environment at a later time.

Swarm Intelligence is an innovative distributed intelligent paradigm for solving optimization problems that originally took its inspiration from the biological examples by swarming, flocking and herding phenomena in vertebrates. Data Mining is an analytic process designed to explore large amounts of data in search of consistent patterns and/or systematic relationships between variables, and then to validate the findings by applying the detected patterns to new subsets of data.

Online Optimization

Whether costs are to be reduced, profits to be maximized, or scarce resources to be used wisely, optimization methods are available to guide decision-making.

In online optimization, the main issue is incomplete data and the scientific challenge: how well can an online algorithm perform? Can one guarantee solution quality, even without knowing all data in advance? Inreal-time optimization there is an additional requirement: decisions have to be computed very fast in relation to the time frame we are considering.

Chapter 7

HERBICIDE-INTERCALATED ZINC LAYERED HYDROXIDE NANOHYBRID FOR A DUAL-GUEST CONTROLLED RELEASE FORMULATION

Mohd Zobir Hussein[1,*], Nor Shazlirah Shazlyn Abdul Rahman[1], Siti H. Sarijo[2] and Zulkarnain Zainal[1,3]

[1] Department of Chemistry, Faculty of Science, Universiti Putra Malaysia, Serdang, Selangor 43400, Malaysia; E-Mails: nsshazlyn@gmail.com (N.S.S.A.R.); zulkar@science.upm.edu. my (Z.Z.)

[2] Faculty of Applied Science, Universiti Teknologi MARA (UiTM), Shah Alam, Selangor 40450, Malaysia; E-Mail: siti_halimah_404@yahoo.com

[3] Advanced Materials and Nanotechnology Laboratory, Institute of Advanced Technology (ITMA), Universiti Putra Malaysia, Serdang, Selangor 43400, Malaysia

* Author to whom correspondence should be addressed; E-Mail: mzobir@science.upm.edu. my; Tel.: +6-03-8946-6801; Fax: +6-03-8943-5380.

ABSTRACT

Herbicides, namely 4-(2,4-dichlorophenoxy) butyrate (DPBA) and 2-(3-chlorophenoxy) propionate (CPPA), were intercalated simultaneously into the interlayers of zinc layered hydroxide (ZLH) by direct reaction of zinc oxide with both anions under aqueous environment to form a new nanohybrid containing both herbicides labeled as ZCDX. Successful intercalation of both anions simultaneously into the interlayer gallery space of ZLH was studied by PXRD, with basal spacing of 28.7 Å and supported by FTIR, TGA/DTG and UV-visible studies. Simultaneous release of both CPPA and DPBA anions into the release media was found to be governed by a pseudo second-order equation. The loading and percentage release of the

DPBA is higher than the CPPA anion, which indicates that the DPBA anion was preferentially intercalated into and released from the ZLH interlayer galleries. This work shows that layered single metal hydroxide, particularly ZLH, is a suitable host for the controlled release formulation of two herbicides simultaneously.

Keywords

4-(2,4-dichlorophenoxy) butyrate; 2-(3-chlorophenoxy) propionate; zinc layered hydroxide; simultaneous release.

1. INTRODUCTION

Nanomaterials such as two-dimensional (2D) nanosheets have recently gained much attention due to their unique physical and chemical properties [1]. Excellent intercalation properties of 2D layered material offer a new scope for developing hybrid materials at nanoscale dimensions or the so-called nanocomposite. This type of material offers a variety of applications in industries and the environment such as anion-exchanger [2], catalysis, delamination, as well as in medical science, and more [3,4].

Layered double hydroxides (LDHs) and hydroxy double salts (HDSs) have been studied extensively and are recognized for their anion-exchangeable properties [5,6]. LDHs having brucite-type layers of mixed metal hydroxide can be expressed by the general formula $[M(II)_{1-x}M(III)_x(OH)_2](A^{-n})_{x/n} \, yH_2O$ where $M(II)$ and $M(III)$ are the divalent and trivalent metal cations, respectively. A^{n-} is the exchangeable anion and y represents the water content of the interlayer region. HDSs are identical to LDHs, only the metal hydroxide inorganic layers are composed of two divalent metal cations [7]. Layered hydroxide salts (LHS) such as zinc layered hydroxide (ZLH) are similar to the HDS structure whose structure is similar to that of brucite; however, the inorganic layers are composed of only one type of metal cation [8] such as Mg^{2+}, Cu^{2+}, Zn^{2+} and Ni^{2+} and can be represented by the general formula, $M(II)(OH)_{2-x}(A^{-n})_{x/n} \, yH_2O$. In this structure, the OH^- anions on the brucite hydroxide layer are substituted by water molecules and counter anions [9–12].

The gallery structure of LDHs is expandable, if the guest anion is larger in size than the interlayer anion present in the LDHs, or the spatial orientation of the guest warrants the expansion. Various guest anions can be intercalated into the LDH's gallery structure, thus a variety of hybrid nano-layered materials can be designed [2,13]. Identical to the LDHs phase, ZLH too can undergo anion-exchange reaction, by substituting the negatively charged organic molecules for the exchangeable interlayer anions in the ZLH lattice to form layered nanohybrids [14,15]. Development of ZLH hybrid materials has shown great potential application in industry such as anticorrosion agent and dye-sensitized solar cells [16,17].

To our knowledge, unlike LDHs, the use of ZLH as matrices [15,18,19], particularly in controlled release (CR) formulations, has rarely been studied. In the agricultural sector, the excess amount of pesticide runoff into the surface and groundwater has led to water pollution [20]. Therefore, the CR formulation can be applied to lower the risks of environmental pollution by reducing the amount of pesticides used for the same activity, thus decreasing the nontarget effects [21]. Additionally, CR formulation is superior to its counterpart and results in a higher yield and better crop quality. Such a formulation also finds use in active agents such as drugs [22,23], vitamins [24,25], herbicides, pesticides and plant growth regulators [26,27] in which the active agents are successfully intercalated into layered materials to produce controlled release formulations.

This paper aimed at the formation of phase-pure, well-ordered ZLH-intercalated nanohybrids by the intercalation of phenoxy herbicides-type active agents, namely 4-(2,4-dichlorophenoxy) butyrate (DPBA) and 2-(3-chlorophenoxy) propionate (CPPA) simultaneously into the ZLH interlayer using a simple direct reaction of zinc oxide (ZnO) with the guest anions in an aqueous environment. In our previous study, we reported the successful intercalation of both herbicides anions into LDHs by an anion-exchange method. Our results showed that the loading and release percentages of DPBA were found to be higher than CPPA [28]. In this work we report the effect of the anion size and their simultaneous controlled release property using zinc layered hydroxide as the host material, which can be employed as a new promising host delivery system for controlled release purpose similar to LDHs.

2. EXPERIMENTAL SECTION

2.1. Synthesis of ZLH Nanohybrid

All solutions were prepared using deionized water. The ZLH nanohybrid was prepared by the direct reaction of ZnO with the guest anions in an aqueous environment. About 1.00 g of pure commercial ZnO was reacted with 100 mL deionized water and mixed with 50 mL aqueous solution of 0.1 M DPBA and 0.1 M CPPA. The solution mixture of ZnO and the anions were stirred for 2 h before aging for 18 h at 70 °C in an oil bath shaker. The slurry was centrifuged, washed with deionized water, dried in an oven for 24 h and kept in a sample bottle for further use and characterization.

2.2. Herbicides Release Study

Simultaneous release of CPPA and DPBA from the ZCDX nanohybrid was studied by adding a 0.4 mg sample into 3.5 mL of 0.001-0.004 mol/L sodium carbonate aqueous solution. The quantity of phenoxyherbicides released into the solution was measured at the preset time at λ_{max} = 220.0 nm and 230.0 nm for CPPA and DPBA, respectively.

2.3. Characterization

Powder XRD patterns were recorded at room temperature with a Shimadzu XRD-6000 using filtered CuK$_\alpha$ radiation (λ = 1.5405 Å) at 40 kV and 30 mA, 2° ·min^{-1}. The FTIR spectra were collected in a Perkin-Elmer 1752X spectrophotometer using the KBr disc method, in the range 400-4000 cm^1. Thermal analyses (TGA-DTG) were performed using a Mettler Toledo instrument at a heating rate of 10 °C min^{-1} between 35 °C and 1000 °C, under nitrogen flow of about 50 mL min^1. A field emission scanning electron microscope (FESEM), Carl Zeiss Supra 40VP model was used to study the surface morphology of the materials.

The percentage loading of the dual herbicides intercalated into the ZCDX nanohybrid was determined using a Perkin Elmer UV-Visible Spectrophotometer, Lambda 35 at λ_{max} = 220.0 nm and 230.0 nm for CPPA and DPBA, respectively. The nanohybrid was dissolved into 2.0 M sodium carbonate aqueous solution so that both anions could exchange with CO_3^{2-} and OH^- anions in the aqueous solution and release completely from the ZLH interlayers. Since the wavelengths of the pure herbicides is relatively close, there is an overlap between the two peaks (Figure 1), hence the data obtained was calculated by solving the simultaneous Equations 1 and 2,

$$a\lambda_1 : A_1 = \varepsilon_{DPBA} \times b \times C_{DPBA} + \varepsilon_{CPPA} \times b \times C_{CPPA} \tag{1}$$

$$at \ \lambda_2 : A_2 = \varepsilon_{DPBA}' \times b \times C_{DPBA} + \varepsilon_{CPPA}' \times b \times C_{CPPA} \tag{2}$$

where λ_1 and λ_2 are the wavelengths for DPBA and CPPA, respectively; A_1 and A_2 indicate the absorbance for the 100 ppm solution containing DPBA and CPPA at λ = 230 nm and λ = 220 nm, respectively, ε is the absorptivity of each anion, and C (mg/L) represents the concentration of 100% release of anions and b is the path length (1 cm). As shown in Equations (1) and (2), the concentration of DPBA and CPPA were obtained by solving the simultaneous equations, using ε_{DPBA} at λ_1 and ε_{DPBA}' at λ_2.

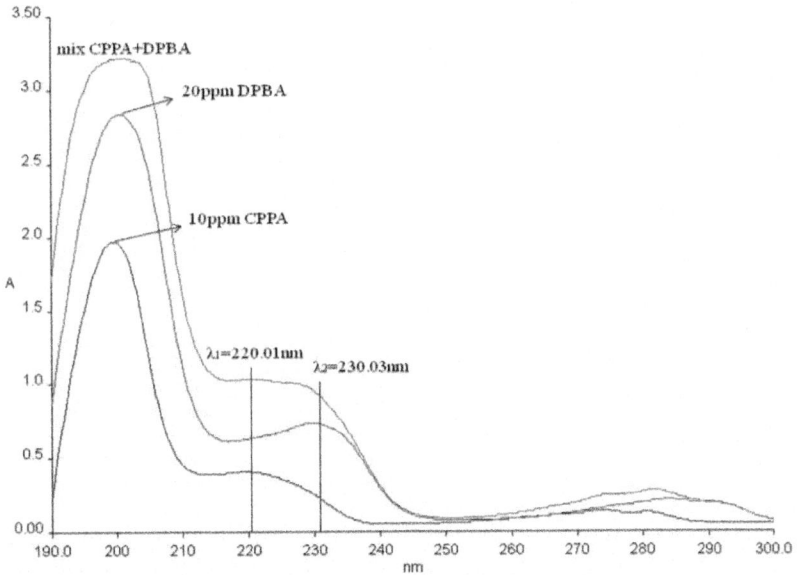

Figure 1. UV-vis spectra of pure anion for 2-(3-chlorophenoxy) propionate (CPPA) and 4-(2,4-dichlorophenoxy) butyrate (DPBA) at 220 and 230 nm, respectively.

3. RESULTS AND DISCUSSION

3.1. PXRD Analysis

The PXRD spectra and basal spacing for the ZLH nanohybrid intercalated with dual herbicides, CPPA and DPBA, labeled ZCDX together with single intercalation of both anions separately via direct reaction with ZnO are shown in Figure 2a.

Several studies has reported on the preparation of LHS by the hydrolysis of divalent metal salts and a metal oxide such as ZnO, followed by the inclusion of guest molecules into the host interlayers [6,7,12]. In our work, a one-step and direct reaction method, which involved dissociation-deposition mechanism, was adopted for the formation of the ZCDX nanohybrid [27,29]. ZnO was first added into the solution containing both anions. As a result it was hydrolyzed and formed $Zn(OH)_2$ on the surface. The dissociation of $Zn(OH)_2$ in the aqueous environment resulted in the release of Zn^{2+} and OH^-. These two species then reacted with CPPA, DPBA and H_2O in the solution to produce the layered ZLH-CPPA-DPBA nanohybrid compound.

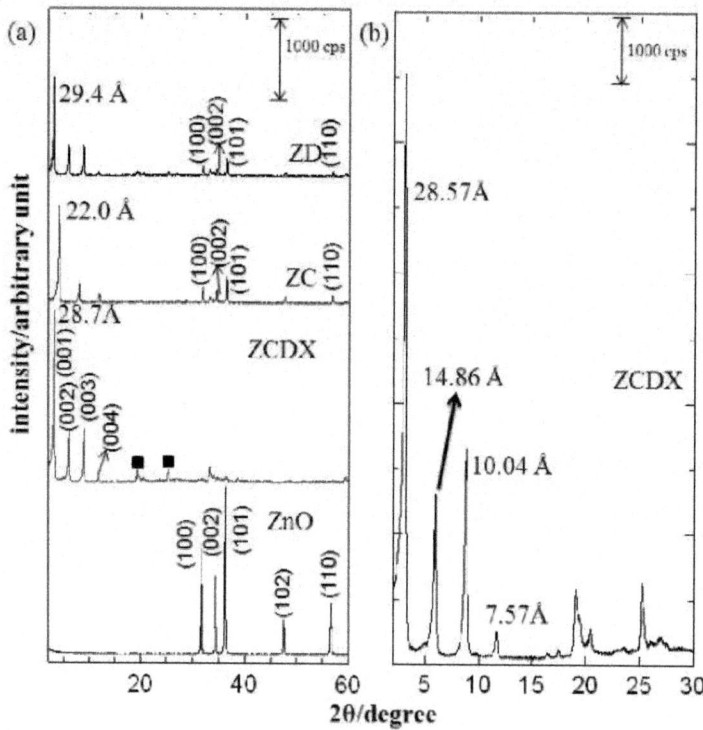

Figure 2. PXRD patterns of pure ZnO, a dual-guest nanohybrid ZCDX prepared at 0.1 M concentration of CPPA and DPBA anion, nanohybrid intercalated with single anion, CPPA (ZC) and DPBA (ZD) at 0.4 M CPPA and 0.2 M DPBA, respectively (a) together with slow scan of ZCDX (b). (■ refers to unknown phase).

Figure 2a shows very sharp and symmetric peaks for ZCDX at the lower 2θ angle due to the intercalation of CPPA and DPBA into the interlayer region. The PXRD patterns of ZCDX prepared at 0.1 M concentration of both anions display a high intensity diffraction peak indicating a pure phase material without any ZnO phase, especially the 100, 002 and 101 reflections [6,10,27]. This shows that a well-ordered nano-layered structure with good crystallinity was obtained at this optimum condition. A slow scan of ZCDX exhibits 4 harmonics at 2θ = 28.57 Å, 14.86 Å, 10.04 Å and 7.57 Å (Figure 2b), resulting in an average basal spacing of 29.67 Å. The obtained basal spacing value is higher than those reported for the intercalation of other type of herbicides into the LDH interlayers [20,26]. ZLH reportedly has larger interspacing than LDH to accommodate a greater number of incoming guest anions of varying sizes, due to its higher charge density [4,14,15]. Thus it is possible to simultaneously intercalate CPPA and DPBA anions into the ZLH interlayers.

3.2. Surface Morphology

The LHS can be described as morphologically having a plate-like structure, as reported elsewhere [10–12,15]. Figure 3 shows the surface morphology of pure commercial ZnO, exhibiting non-uniform granular structure with a particle size around 2 μm [27]. On simultaneous intercalation of the CPPA and DPBA anions into the ZLH interlayers, the ZnO phase changes in structure and was transformed into a mixture of granular and fiber-like structure. This shows that the transformation of ZnO into a nanohybrid produces surface morphology transformation.

Figure 3. FESEM micrographs of ZnO (**a**) and zinc layered hydroxide (ZLH) nanohybrid (**b**) synthesized by direct reaction of ZnO at 10,000× magnification. The higher magnification of ZnO and ZLH nanohybrid are given in the insets.

3.3. Spatial Orientation of the Dual Herbicides in the ZLH Interlayers

Figure 5 shows the proposed arrangement of the dual herbicides, CPPA and DPBA within the ZLH interlayer region. This is based on the PXRD data and the molecular size of both anions calculated using Chem3d Ultra 9.0 software [28] as shown in Figure 4. The ZLH is composed of inorganic layers with octahedral coordinated zinc cations, of which 1/4 are displaced out of the layer, leaving an empty octahedral site. Tetrahedrally coordinated Zn^{2+} located at the top and bottom of the octahedral sheet [4,5,7,11,17] are noted. Based on this structure, the inorganic layer thickness is 4.8 Å, including 2.6 Å for each zinc tetrahedron [9] and the basal spacing of the ZCDX nanohybrid is 29.67 Å. Therefore, the expected gallery height occupied by the two herbicides in the interlayer space of ZLH is 19.67 Å.

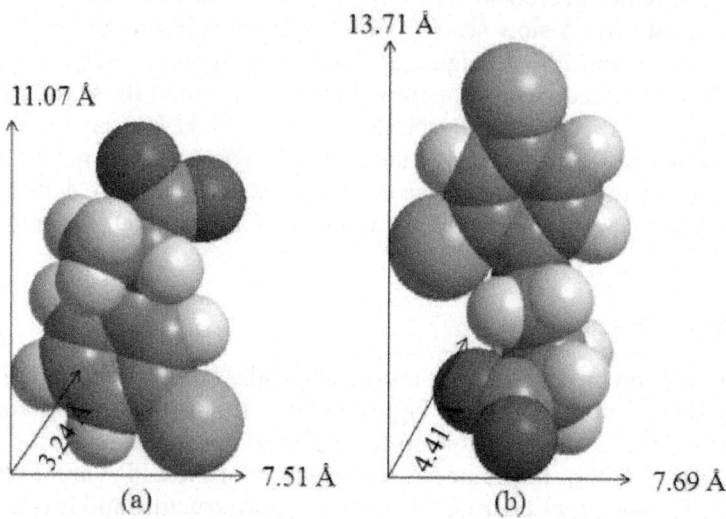

Figure 4. 3D-Molecular structures of (**a**) 2-(3-chlorophenoxy)propionate; (CPPA) and (**b**) 4-(2,4-dichlorophenoxy)butyrate (DPBA).

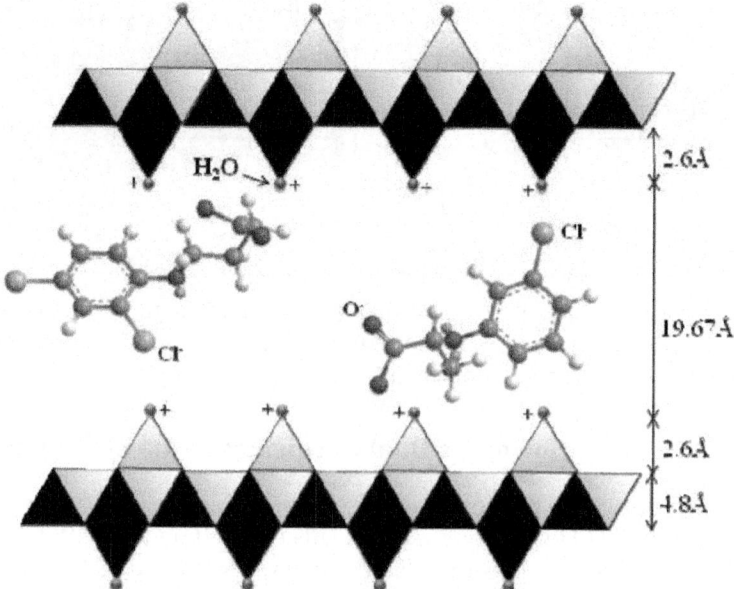

Figure 5. Proposed spatial orientation of CPPA and DPBA in the ZLH inorganic interlayers.

Considering several factors such as charge density of the layer, anion dimension and assuming that the layer structure remains intact after the intercalation of both anions [9,15], then the CPPA and DPBA have to orient themselves in a monolayer arrangement similar to those reported for salicylic acid with an interlayer spacing of 16.0 Å [14]. The oxygen atom in the carboxylate groups and also chlorine atoms attached to the benzene ring are directly bonded to the ZLH layers through hydrogen bonding and electrostatic interaction. Such an arrangement leads to stronger interaction between the host layers and both the anions which cause the increase in thermal stability of the layered structure [3,11,14,24].

3.4. Thermal Studies

The thermal decomposition behavior of ZnO and ZCDX nanohybrid was investigated using TGA-DTG studies. It is known that the LHS thermograms are generally described by two thermal events which are similar to that for organo-LDHs [5,12]. The first endothermic peak is observed with the temperature range of room temperature to nearly 200 °C, commonly assigned to the loss of adsorbed and structural water molecules of the ZLH layers. At approximately 300 °C, the weight loss is attributed to simultaneous decomposition of anions and dehydroxylation of the zinc hydroxide layer [1,4,8,9]. As shown in Figure 6, no thermal decomposition was observed for pure ZnO, which proved that it is a thermally stable compound. CPPA and DPBA anions show an endothermic peak at temperature maxima of 228.8 °C and 274.1 °C with weight loss of 97.8% and 99.6%, respectively.

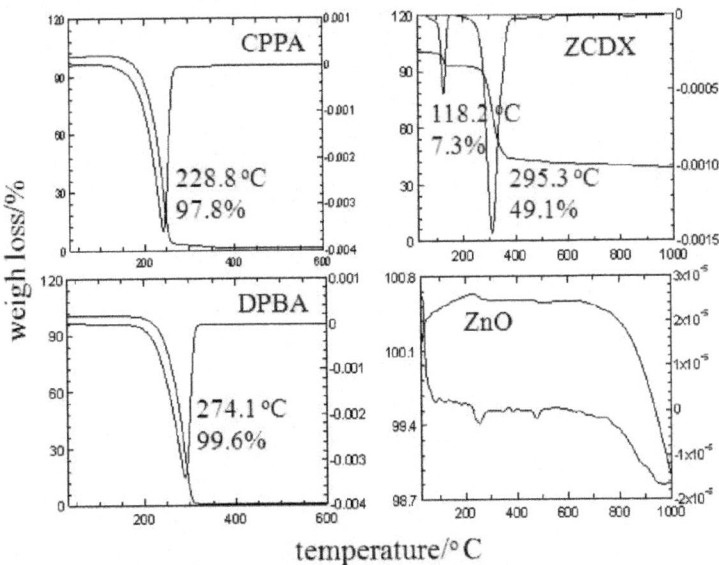

Figure 6. TGA-DTG thermograms of CPPA, DPBA, dual-guest nanohybrid ZCDX and ZnO.

The TG curve of ZCDX shows two major weight losses at temperature maxima of 118.2 °C and 295.3 °C. The first stage of weight loss with 7.3% is ascribed to the removal of the surface and intercalated structure of water molecules, leading to the decomposition for dual herbicides of the nanohybrid at 295.3 °C with 49.1% weight loss. This shows that the ZCDX nanohybrid is thermally more stable than CPPA and DPBA in their salt form [9,26]. DPBA has higher thermal stability since it decomposed at a higher temperature compared to the CPPA anion. We assumed that the DPBA anion has higher stability and stronger interaction with the ZLH interlayer compared with the CPPA due to the two chlorine atoms attached to the benzene ring, which may induce stronger polar interaction with ZLH layers based on the proposed orientations of both anions in the interlayers as mentioned earlier.

3.5. FTIR Analysis

The presence of the two herbicides in the ZLH interlayers was also elucidated by FTIR spectroscopy, which supported the PXRD results [2]. Figure 7 compares the FTIR spectrum of pure ZnO, intercalated compound, ZCDX with CPPA and DPBA anions. The sharp and intense peak of ZnO at the lower wavenumber range, 385 cm^{-1} is due to the Zn–O sublattice stretching vibration [8,27]. Similar absorption bands could also be observed for the CPPA and DPBA anions, as they possess the same functional groups. Broad peaks at 2995 and 2969 cm^{-1} are attributed to the OH stretching vibration of COOH for CPPA and DPBA anions, respectively. Intense peaks at around 1700 cm^{-1}, for CPPA (1705 cm^{-1}) and DPBA (1714 cm^{-1}) are ascribed to C=O stretching vibration of the protonated COOH group of herbicides. Strong bands at 1471 and 1467 cm^{-1} are due to the stretching vibration of the aromatic ring, C=C and bands at around 1200 cm^{-1} are due to C–O–C antisymmetric and symmetric stretching.

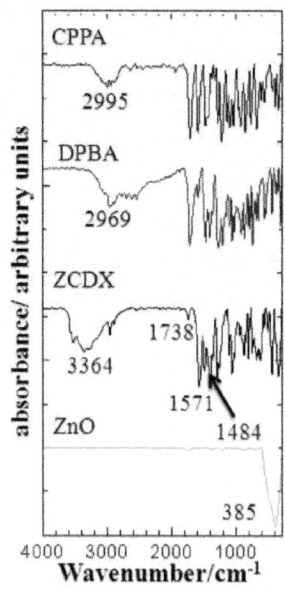

Figure 7. FTIR spectra of the guest anions CPPA, DPBA, the nanohybrid ZCDX and ZnO.

As expected, ZCDX has an absorption spectrum similar to the herbicide anions. However, some bands are slightly shifted due to the interaction of both the anions and host layers. Weak bands at around 1700 and 1400 cm^{-1} are due to the stretching vibration of the C=O of the COOH group and C=C stretching of the aromatic ring, as mentioned earlier, indicating that the two herbicides were successfully intercalated into the ZLH interlayers [20]. A strong peak at 1571 cm^{-1} is due to the antisymmetric and symmetric carboxylate stretching of the anion. At the lower wavenumber, the disappearance of the broad ZnO peak in the nanohybrid confirmed that all the ZnO had reacted with the CPPA and DPBA, resulting in the formation of the nanohybrid, which concurs with the PXRD data.

3.6. Herbicides Release Properties

The loading percentages of CPPA and DPBA were analyzed using a UV-Vis spectrophotometer and the data were calculated using simultaneous equations, as stated earlier. The data show that the ZCDX nanohybrid is composed of CPPA and DPBA with percentage contribution of 16.22% and 83.78%, respectively. The percentage represents the ratio of CPPA and DPBA intercalated into ZLH. The simulation data of both single intercalations of ZC and ZD nanohybrids at different percentages (Figure 8) shows that a good fit was obtained when the PXRD patterns were obtained using 15% CPPA and 85% DPBA contribution. The high content of the DPBA anion reflects the higher abundance of DPBA in the interlayer gallery space of ZLH compared with the CPPA and this in turn will influence the release behavior.

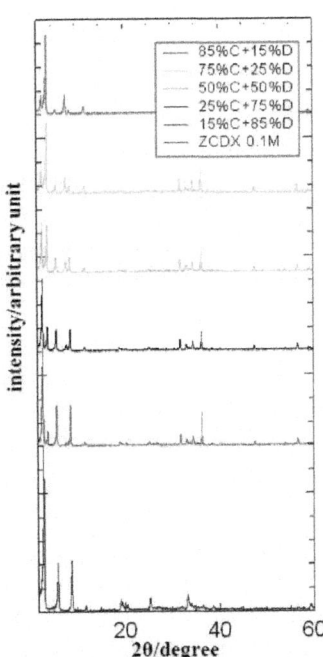

Figure 8. PXRD patterns of ZCDX nanohybrid and simulated by addition of PXRD patterns of single intercalation nanohybrid ZC and ZD at different percentages.

The simultaneous release of CPPA and DPBA from the nanohybrid lamellae into the release media was performed in various concentrations (0.001–0.004 mol/L) of sodium carbonate aqueous solution. The affinity of CO_3^{2-} and OH^- anions in the aqueous solution is higher than the CPPA and DPBA towards the ZLH interlayers; therefore, both herbicides anions can be ion-exchanged with the carbonate [2,20,22,25] and the release of CPPA and DPBA was determined at λ_{max} = 220 and 230 nm, respectively. Generally, the release rate is initially rapid, and then slows until it reaches equilibrium. The percentage release of CPPA and DPBA increased when the ZCDX was placed in a higher concentration of sodium carbonate, as shown in Figure 9a–c. Maximum release was achieved at about 2000 min (approximately after a day) into 0.004 mol/L sodium carbonate. Taking the intercalated amount of the CPPA and DPBA in ZCDX as 100% respectively, the maximum accumulated release of CPPA was found to be 53.3% which is less than DPBA with maximum saturated release of 68%.

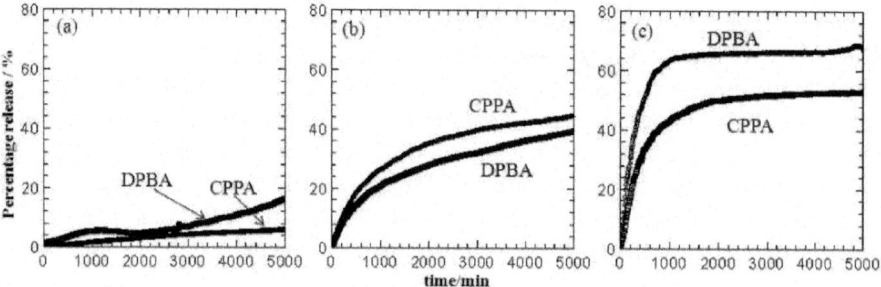

Figure 9. Release profiles of CPPA and DPBA from the dual-guest nanohybrid, ZCDX into various concentrations of Na_2CO_3 aqueous solution: 0.001 mol/L (**a**), 0.002 mol/L (**b**), and 0.004 mol/L (**c**).

Previous work revealed that release behaviors of different drugs from the same host are quite different [24]. Such behavior could be related to several factors such as the guest size, preparation method of the nanohybrid, anion affinity in the release medium, particle size, packing density and chemical interactions between the host and guest [15,17,24–26]. Our results clearly proved that the release of the CPPA and DPBA depends on the anionic size as well as on the host-guest and/or guest-guest interaction between the host interlayers. The high uptake of DPBA anion is due to its larger size and stronger interaction within the high-charged density of the ZLH interlayers when two Cl^- atoms bonded with Zn^{2+} ions compared to only one Cl^- atom in the CPPA [28], as proposed in Figure 5. The abundance of DPBA anion (high loading percentage) makes it easier to be released in high concentration sodium carbonate solution.

The data of the cumulative release of the guests were fitted to several kinetic models and plotted in Figure 10. Among the models used in this work, the correlation coefficient (r^2) for the pseudo second-order kinetic gave the best fit for both CPPA and DPBA, as shown in Table 1. The half-life, $t_{1/2}$ values for the DPBA and CPPA which are 286 and 543 min, respectively, confirmed that the former is easier to release than the latter from the nanohybrid, when both anions are present simultaneously. In most of controlled release formulation, a large amount of anions (in this case, DPBA and CPPA) were released at the very beginning of the release time before the release rate reaches a stable profile. This is due to the swelling of the porous structure followed by the dissolution of the anions. This phenomenon is referred to as "burst release" [23], which happens in a very short period of time compared to the entire release process, resulting in the deviation from linearity of the pseudo-second order fit at the beginning of the release process.

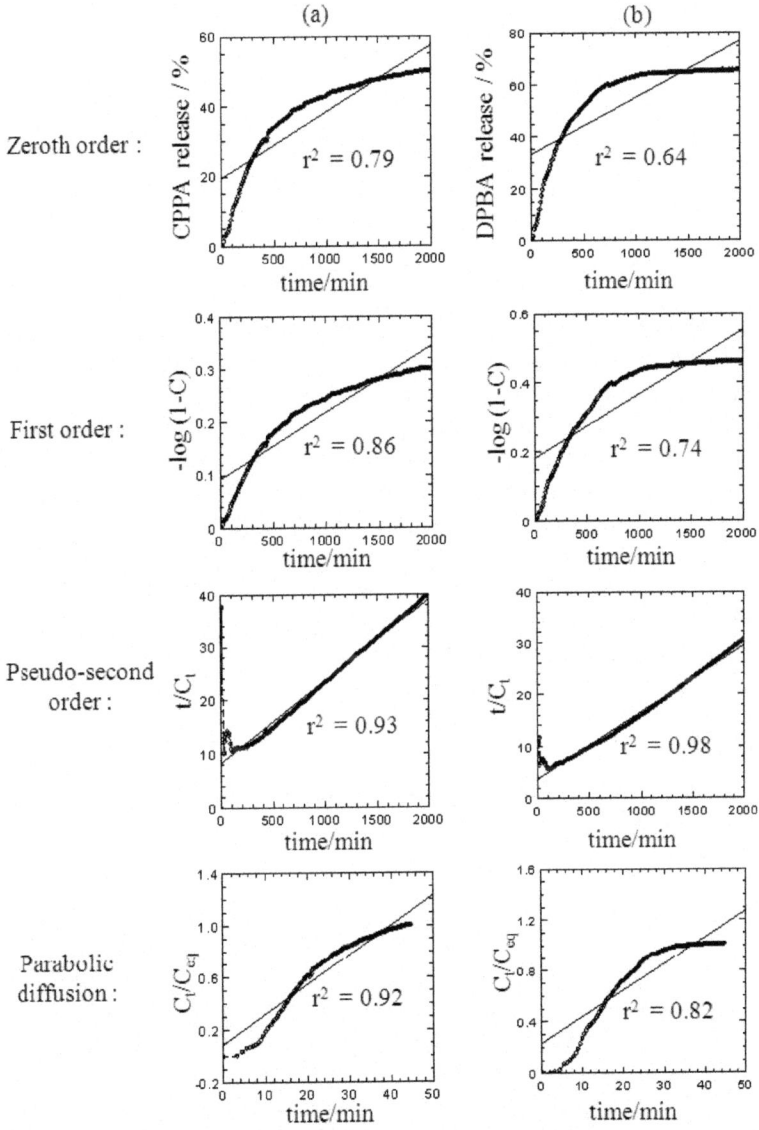

Figure 10. Fitting the release data of CPPA (**a**) and DPBA (**b**) from ZCDX nanohybrid into 0.004 mol/L Na_2CO_3 solution to the zero-, first-, pseudo second-order kinetics and parabolic diffusion model.

Table 1. Correlation coefficient, r^2 values obtained from fitting the data of simultaneous release of CPPA and DPBA to zero-, first- and pseudo-second order kinetics. The half-life, $t_{1/2}$, k and c values for pseudo-second order kinetics are also shown.

Models	Zeroth Order	First Order	Parabolic Diffusion	Pseudo-Second Order			
Anion		r^2		r^2	$t_{1/2}$ (0–3000 min)	k ($\times 10^{-5}$) (Lmg^{-1}min^{-1})	c
CPPA	0.79	0.86	0.92	0.93	543	2.82	8.31
DPBA	0.64	0.74	0.82	0.98	286	4.51	3.69

equations

Zeroth order: $C = kt + c$ Pseudo-second order: $t/C_t = 1/k_2 C_{eq}^2 + (1/q_e) t$

First order: $-\log(1-C) = kt + c$ Parabolic diffusion: $C_t/C_{eq} = c + kt^{0.5}$

C_{eq} = concentration of anion at equilibrium, C_o = initial concentration of the anions, C_t = concentration of anion at time t, C = percentage release of anion, c = a constant.

4. CONCLUSIONS

In conclusion, the monophasic, well-ordered ZCDX nanohybrid containing two herbicides, CPPA and DPBA, was found to be composed of a higher loading of DPBA compared to CPPA between the ZLH inorganic interlayers with percentage contribution of 83.78% and 16.22%, respectively. Simultaneous release of CPPA and DPBA from ZCDX into sodium carbonate aqueous solution was found to be controlled by pseudo second-order kinetics. The release rate of both CPPA and DPBA was found to be different, suggesting that the anionic guest molecules sizes and the interactions between the host and guest could control the release kinetics. The present results demonstrate the potential application of a layered single metal hydroxide, particularly zinc layered hydroxide as the host for the preparation of a nanohybrid compound with tunable controlled release property containing two herbicides simultaneously.

Acknowledgment

Funding for this research was provided by the Fundamental Research Grant Scheme (FRGS) under GRANT No. 02-11-08-615FR and a Graduate Research Fellowship (GRF) by Universiti Putra Malaysia for NSSAR.

REFERENCES

1. Miao, J.; Xue, M.; Itoh, H.; Feng, Q. Hydrothermal synthesis of layered hydroxide zinc benzoate compounds and their exfoliation reactions *J. Mater. Chem.* **2006**, *16*, 474–480.

2. Lv, L.; Sun, P.; Gu, Z.; Du, H.; Pang, X.; Tao, X.; Xu, R.; Xu, L. Removal of chloride ion from aqueous solution by ZnAl-NO$_3$ layered double hydroxides as anion-exchanger. *J. Hazmat.* **2009**, *161*, 1444–1449.

3. Nalawade, P.; Aware, B.; Kadam, V.J.; Hirlekar, R.S. Layered double hydroxides: A review. *J. Sci. Ind. Res.* **2009**, *68*, 267–272.

4. Kasai, A.; Fujihara, S. Layered single-metal hydroxide/ethylene glycol as a new class of hybrid material. *Inorg. Chem.* **2006**, *45*, 415–418.

5. Newman, S.P.; Jones, W. Comparative study of some layered hydroxide salts containing exchangeable interlayer anions. *J. Solid State Chem.* **1999**, *148*, 26–40.

6. Morioka, H.; Tagaya, H.; Kasaru, M.; Kadokawa, J.; Chiba, K. Effects of zinc on the new preparation method of hydroxyl double salts. *Inorg. Chem.* **1999**, *38*, 4211–4216.

7. Kandare, E.; Hossenlopp, J.M. Hydroxyl double salt anion exchange kinetics: Effects of precursor structure and anion size. *J. Phys. Chem.* **2005**, *109*, 8469–8475.

8. Hussein, M.Z.; Ghotbi, M.Y.; Yahaya, A.H.; Rahman, M.Z.A. Synthesis and characterization of (zinc-layered-gallate) nanohybrid using structural memory effect. *Mater. Chem. Phys.* **2009**, *113*, 491–496.

9. Marangoni, R.; Ramos, L.P.; Wypych, F. New multifunctional materials obtained by the intercalation of anionic dyes into layered zinc hydroxide nitrate followed by dispersion into Poly (vinyl alcohol) (PVA). *J. Colloid Interface Sci.* **2009**, *330*, 303–309.

10. Altuntasoglu, O.; Matsuda, Y.; Ida, S.; Matsumoto, Y. Syntheses of zinc oxide and zinc hydroxide single. *Nanosheets Chem. Mater.* **2010**, *22*, 3158–3164.

11. Zhao, L.; Miao, J.; Wang, H.; Ishikawa, Y.; Feng, Q. Synthesis and exfoliation of layered hydroxide zinc aminobenzoate compounds. *J. Ceram. Soc. Jpn.* **2009**, *117*, 1115–1119.

12. Arizaga, G.G.C.; Satyanarayana, K.G.; Wypych, F. Layered hydroxide salts: Synthesis, properties and potential applications. *Solid State Ion.* **2007**, *178*, 1143–1162.

13. Kuk, W.K.; Huh, Y.D. Preferential intercalation of organic anions into layered double hydroxide. *Bull. Korean Chem. Soc.* **1998**, *19*, 1032–1036.

14. Hwang, S.H.; Han, Y.S.; Choy, J.H. Intercalation of functional organic molecules with pharmaceutical, cosmeceutical and nutraceutical functions into layered double hydroxides and zinc basic salts. *Bull. Korean Chem. Soc.* **2001**, *22*, 1019–1022.

15. Yang, J.H.; Han, Y.S.; Park, M.; Park, T.; Hwang, S.J.; Choy, J.H. New inorganic-based drug delivery system of indole-3-acetic acid-layered metal hydroxide nanohybrids with controlled release rate. *Chem. Mater.* **2007**, *19*, 2679–2685.

16. Rocca, E.; Caillet, C.; Mesbah, A.; Francois, M.; Steinmetz, J. Intercalation in zinc-layered hydroxide: Zinc ydroxyheptanoate used as protective material on zinc. *Chem. Mater.* **2006**, *18*, 6186–6193.

17. Demel, J.; Kubát, P.; Jirka, I.; Kovář, P.; Pospíšil, M.; Lang, K. *J. Phys. Chem.* **2010**, *114*, 16321–16328.

18. Bull, R.M.R.; Markland, C.; Williams, G.R.; O'Hare, D. Hydroxy double salts as versatile storage and delivery matrices. *J. Mater. Chem.* **2011**, *21*, 1822–1828.

19. Hussein, M.Z.; Al Ali, S.H.; Zainal, Z.; Hakim, M.N. Development of antiproliferative nanohybrid compound with controlled release property using ellagic acid as the active agent. *Int. J. Nanomed.* **2011**, *6*, 1373–1383.

20. Cardoso, L.P.; Celis, R.; Cornejo, J.; Valim, J.B. Layered double hydroxides as supports for the sow release of acid herbicides. *J. Agric. Food Chem.* **2006**, *54*, 5968–5975.

21. Lewis, D.; Cowsar, D. Principles of Controlled Release Pesticides. In *Controlled Release Pesticides*; Scher, H.B., Ed.; Acs Symposium Series 57 American Chemical Society: Washington, DC, USA, 1977; pp. 1–14.

22. Li, B.; He, J.; Evans, G.; Duan, X. Inorganic layered double hydroxides as a drug delivery system-intercalation and in vitro release of fenbufen. *Appl. Clay Sci.* **2004**, *27*, 199–207.

23. Huang, X.; Brazel, C.S. On the importance and mechanisms of burst release in matrix-controlled drug delivery systems. *J. Control. Release* **2001**, *73*, 121–136.

24. Khan, A.I.; Ragavan, A.; Fong, B.; Markland, C.; O'Brien, M.; Dunbar, T.G.; Williams, G.R.; O'hare, D. Recent developments in the used of layered double hydroxides as host materials for the storage and triggered release of functional anions. *Ind. Eng. Chem. Res.* **2009**, *48*, 10196–10205.

25. Gasser, M.S. Inorganic layered double hydroxides as ascorbic acid (vitamin c) delivery system—Intercalation and their controlled release properties. *Colloid Surf.* **2009**, *73*, 103–109.

26. Sarijo, S.H.; Hussein, M.Z.; Yahaya, A.H; Zainal, Z.; Yarmo, M.A. Synthesis of phenoxyherbicides-intercalated layered double hydroxide nanohybrids and their controlled release property. *Curr. Nanosci.* **2010**, *6*, 199–205.

27. Hussein, M.Z.; Hashim, N.; Yahaya, A.H.; Zainal, Z. Synthesis and characterization of [4-(2,4-dichlorophenoxybutyrate)-zinc layered hydroxide] nanohybrid. *Solid State Sci.* **2010**, *12*, 770–775.

28. Hussein, M.Z.; Rahman, N.S.S.A.; Sarijo, S.H.; Zainal, Z. Synthesis of a monophasic nanohybrid for a controlled release formulation of two active agents simultaneously. *Appl. Clay Sci.* **2012**, *58*, 60–66.

29. Xingfu, Z.; Zhaolin, H.; Yiqun, F.; Su, C.; Weiping, D.; Nanping, X. Microspheric organization of multilayered ZnO nanosheets with hierarchically porous structures. *J. Phys. Chem.* **2008**, *112*, 11722–11728.

INDEX